ROOTS OF NEURO-LINGUISTIC PROGRAMMING

BY

Robert Brian Dilts

Dilts Strategy Group
P.O. Box 67448
Scotts Valley, California 95067
Phone: +1 (831) 438-8314
E-Mail: info@diltsstrategygroup.com
Homepage: http://www.diltsstrategygroup.com

©Copyright 1983 by Robert B. Dilts and Dilts Strategy Group. All rights reserved. This book or parts thereof may not be reproduced in any form without written permission of the Publisher.

Library of Congress Card Number 96-075856
I.S.B.N. 978-1-947629-20-2

PART I: ROOTS OF NEURO-LINGUISTIC PROGRAMMING

(1976)

BY

Robert Brian Dilts

INTRODUCTION
(NOVEMBER, 1976)

The goal of Neuro-Linguistic Programming (NLP) is to *integrate the macroscopic information* about human behavior and experience available to each of us through our *sensory experience* with *the unobservable microscopic information* of the *neuro-physiology of behavior and experience* into a useful *cybernetic model*. I believe such an integration is essential if we ever hope to understand or utilize the properties governing the complex human processes of learning, memory, communication, choice, and motivation and how these affect the social and ecological environments of human beings.

Neuro-Linguistic Programming was founded by behavioral modelers John Grinder and Richard Bandler to analyze and explore the patterns governing such complex processes of human behavior. The basic premise of NLP is that there is a redundancy between the *observable macroscopic patterns of human behavior* (for example, linguistic and paralinguistic phenomena, eye movements, hand and body position, and other types of performance distinctions) and *patterns of the underlying neural activity governing this behavior.*

For example, we have probably all had the experience of noticing, in the course of some conversation, another individual, in response to some question, slow down the tempo of his or her voice while trying to recall the requested information, then suddenly flick his/her eyes up and to the left, breaking perceptual contact momentarily. This movement is generally followed by an increase in tempo as the individual describes or talks about some object or event, usually using language indicative of some sort of visual representation. One might say *that this segment of behavior is indicative* of the *access of some internal visual image* (that is, activity in the visual cortex of the brain separate from the visual information being provided by the individual's immediate sensory experience).

Figure 1. Accessing visual information.

Similarly, if one observes another individual or that same individual orient his or her head down and to the left in response to some question, put a hand over his/her mouth, and mutter "hmmmmm," eventually responding with some utterance describing what that individual has *heard* about some object or event, one might conclude that this segment of behavior is indicative of some neural process that is qualitatively different from that indicated by the behavioral segment previously described. If one can determine the correlation between these segments and the individual's internal experience and patterns (consistency and combinations) involving their use, one might be able to make *assumptions and predictions concerning the personality and behavior of another individual and act on this knowledge.*

The previous two examples are simple and relatively meaningless when presented in the absence of other contextual information concerning the entire communicational sequence. And, indeed, Neuro-Linguistic Programming involves an examination of the entire system of feedback and response, both behaviorally and biologically, occurring in the interactions between (a) a human being and him/herself, (b) a human being and other human beings, and (c) a human being and his/her environment.

Neuro-Linguistic Programming is a process and at the same time a model of a process (as are most scientific and cybernetic models). All of the information, laws, patterns, and conclusions made by the model about past experience are only important or useful in *the way that they relate to one's immediate ongoing experience.* This is especially important to keep in mind when one is dealing with another cybernetic system (in this case, a human being) that is subject to many levels of change, depending on its interactions with itself, its social environment, and its ecological environment.

The following is a *basic outline of the Neuro-Linguistic Programming process:*

A. Train and expand one's own sensory awareness of other human beings to:

1. observe and identify meaningful patterns of behavior which are systematic, recurrent, and a part of everyone's sensory experience.

2. notice what responses one's own choice of behavior elicits in oneself and other human beings (and vice versa).

B. Utilize the information one gathers through these observations to determine:

1. the representational distinctions human beings can make about their internal and external experience, i.e., their ability to see/visualize, hear/verbalize, etc.

2. patterns involving the combination and connections of neural networks (sensory representation) underlying behavioral processes.

3. how these distinctions and patterns affect the strategies people use to organize, make sense of, and communicate about their sensory experience and internal maps.

4. how these distinctions, patterns, and strategies may be utilized to understand and promote the processes of, among other things, learning, communication, motivation, and choice in human beings within the individual and his/her social and ecological environments.

Roots of Neuro-Linguistic Programming

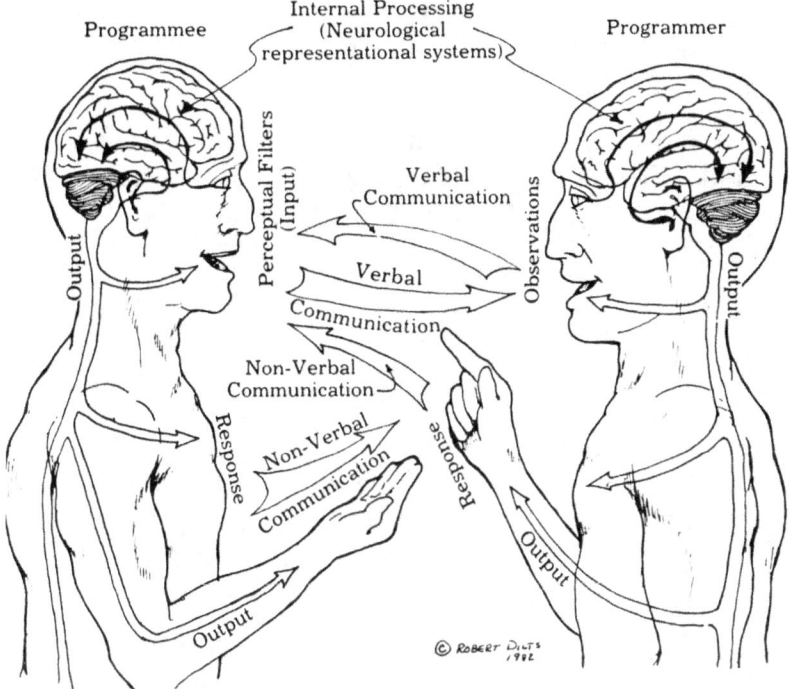

Figure 2. The Neuro-Linguistic Programming process.

The mechanics of this process may be generalized into the following basic procedure:

a. An outcome is identified that is mutually acceptable to the facilitator (programmer) and the client (programmee). Explicit criteria for the successful achievement of the outcome are delineated.

b. One individual (the programmer) generates a communication, in the form of verbal and non-verbal behavior, in an attempt to direct or propel the programmee to the desired outcome. This communication elicits a response in the form of some access of information and return communication on the part of the other individual, the programmee. The interaction will take place on both the verbal and nonverbal level of response.

c. The two individuals may then work together to make distinctions in the programmee's experience and response, both internally and externally:

1. via the programmer's perception of the programmee's external behavior within the specific contextual setting.

2. via explicit discussion and questioning about the programmee's internal experience.

d. The distinctions may then be classified and segmented into a formula that traces the programmee's behavior in terms of relevant *input,* mental *processing* and behavioral *output*.

e. The information provided by these patterns and distinctions is utilized to make decisions and predictions about the individual's current and future behavior, and to devise more effective strategies the programmee can use to achieve desired outcomes.

f. The programmer then varies his/her behavior in accordance with this information, and the process repeats itself until the outcome is reached.

Any lack of clarity in this procedure due to the formality of this description will hopefully be cleared up in the course of these papers (specifically, Part III - Applications of Neuro-Linguistic Programming to Therapy). I intend to proceed as follows:

1) A more explicit discussion of models and modeling—specifically, cybernetic models.

2) Application of cybernetics to a theory of the function of the brain.

3) Some implications of this theory and how it relates to the redundancy between internal experience and external behavior.

4) Discussion of Neuro-Linguistic Programming in relation to these implications.

5) Uses and applications of Neuro-Linguistic Programming as a tool for communication, the achievement of excellence and individual and social change.

The achievement of this ambitious goal involves three manuscripts. Each part has been written to build on the information provided by the previous section but may also be read with little or no reference to the others.

Part I. *Roots of NLP* is a highly technical exploration of the relationship between information theory, neurology and cybernetics as background for the NLP model.

Part II. *EEG and Representational Systems* describes a brain model and research experiment supporting the basic tenets of NLP.

Part III. *Applications of Neuro-Linguistic Programming to Therapy* gives an overview of the applications of NLP in the therapeutic contexts with examples of specific uses of the techniques.

TABLE OF CONTENTS FOR PART I

	Page
I. The Structure of Experience	14
1.1 The Map is Not the Territory	14
A. Forms of Experience	14
1. Immediate Sensory Experience	15
2. Experience of Experience	15
Meta-Section: Consciousness	17
3. Experience About Experience	19
B. The Structure of Epistemology	19
1.2 Cybernetics	20
A. Cybernetic Explanation	21
B. Units of Meaning	21
1. Information and Difference	22
2. Dimensionality and Chunking of Information	23
a. Classification of Difference	23
1) Hierarchies	23
2) Logical Typing	24
3. Redundancy and the Coding of Information	25
a. Coding	25
1) causal and correlative	26
2) analogic	26
3) digital	27

4) iconic	27
5) ostensive	28
6) evolutionary	28
7) holographic	29
b. Entropy	29
c. Discontinuities and Catastrophies	30
C. Patterning	32
1.3 Properties of a Cybernetic System	33
A. Units of Difference and Information	34
B. Classification and Chunking of Information and Difference	36
1. Dimensionality and Measurement	39
(1) Spatial Relations	39
(2) Temporal Relations	39
(3) Causal Relations	39
a. Descriptive Parameters	40
(a) Signal-To-Noise Ratio	40
(b) Simultaneous or Sequential Expression	40
(c) Afferent versus Efferent	41
2. Classification and Interaction	41
a. Properties of Interaction	43
(a) Differences of the same logical type interacting with each other at the same hierarchic or logical level	44
(b) Differences of different logical types interacting at the same logical level	46
(c) Differences of the same or different logical type interacting at different levels	49
b. Feedback	53
(a) Feedback between similar types on the same level	53

(b) Feedback between different types on the same level	54
(c) Feedback between different types on different levels	55
(d) Feedback between similar logical types on different levels	57
c. Coding of Information	58
II. Towards a Cybernetic Theory of Brain Function	58
2.1 Some previous theories of brain function	60
A. Generation of New Connections	60
B. Cell Assemblies	61
C. Causal Loops	62
D. Digital Circuitry	63
2.2 The Statistical Brain	64
2.3 The Cybernetic Brain	68
Appendix A: Electrostatic versus Electromagnetic Field	70
a. Electrostatic field	70
b. Electromagnetic field	70
1. Field effect	72
Bibliography for Part I	76

I. The Structure of Experience

Neuro-Linguistic Programming is a model of the structure of our subjective experience and how that experience influences our behavior. As such, NLP can be considered an *"epistemology"* of experience. Stated simply, an epistemology is "knowledge about knowledge;" a system which deals with the structure and form of the brain's processes, not the content.

1.1 The Map is not the Territory

Epistemological models such as NLP are unique models in that, while they are models about our experience, by the very act of thinking about such models they also become a part of our experience. Because of this unique position, I would like the reader to keep in mind that this monograph is not an attempt to describe *reality*, but rather, it is an attempt to elucidate the way in which we experience reality. Its value does not lie in the accuracy with which it describes reality but rather in its ability to reorganize the way in which you experience and respond to your own model of the world.

A. Forms of Experience

To avoid unleashing and getting caught up in the maze of tautologies that accompanies such a discussion, I would like to begin by dividing this very general and confusing concept, "experience," into three different logical types that I believe are important for the purposes of talking about a behavioral and physiological epistemology.

1. Immediate Sensory Experience.

One might consider immediate sensory experience to be synonomous with "perception." It is the sum total of information that we receive about the "territory" through our sense organs at any particular point in time: the interaction between the firing of the receptors and neurons in an individual's auditory, visual, somatosensory, proprioceptive, gustatory, and olfactory organs and that individual's brain (i.e., hearing, seeing, feeling, tasting and smelling). There are two immediate considerations that come to mind about the process of immediate sensory experience, or indeed any discussion of the process of experience:

a. The limits or degree to which one can be conscious or aware of one's immediate sensory experience, and the effects of that limitation.

b. The amount of conscious or unconscious perceptual analysis that can take place at the level of immediate sensory experience.

To be able to answer these questions in the most complete and graceful manner we must first proceed to a discussion of the next logical type of experience:

2. Experience of Experience.

This level of experience involves the mental maps or models we make to organize and respond to our immediate sensory perceptions. Neurologically these are representations resulting from the stimulation of networks of cortical cells and neurons that form complex branching and interconnecting chains of causation in the brain. It is the complex and overlapping interconnection of cells in the cortex that make this type of experience qualitatively different from immediate sensory experience.

There are two parts to this phenomenon, however, that must be discussed and sorted:

a. The *form* or process by which this experience takes place.

b. The *content* representations that result from that process.

All of the different names that we give this type of experience (memory, thought, learning, and so on) can similarly be thought to be composed of these two parts.

Considering the content aspects first, we can see that representations can form various hierarchic structures: there can be thoughts about thoughts, representations of representations, and so on. The result is a phenomenology of experience based on the different combinations, (and combinations of combinations) of sensory information, i.e., we can identify objects, relationships, events, processes, systems, and classes of these phenomena, that we consider to be qualitatively different.

In addition to these maps, there is another set of representations, generally considered to be unique to human beings, that are the result of the process of *natural language.* Language is a symbolic representation of our sensory representations; "meta" representation so to speak. Words, however, have no meaning in and of themselves and can only be understood by an individual in relation to his/her own sensory model of the world. For instance, if someone is describing a movie that they have seen they may be seeing a remembered image in their mind's eye. Since you cannot see inside their head, however, you must translate their words into a mental picture of your own, which may or may not be similar to theirs.

All forms of representation are the result of the process of combining sensory information, a process that has structure.

Meta-Section: Consciousness

In any discussion of experience or the process of experience it is important to make the distinction between *conscious* and *unconscious* experience. Because a large part of the later sections of this paper will be dealing with this distinction, I will limit my discussion here to two basic points:

1) Consciousness is a secondary process.

2) Learning is not limited by conscious awareness.

In his book *Physiological Psychology* (1970) P.M. Milner points out:

> "It seems that the classifying powers of the higher levels of perceptual system must be called upon to process the input before it is fed into the arousal system. This finding introduces the interesting idea that perceptual analysis takes place quite normally in the absence of attention or while attention is directed to different aspects of the perceptual field.
>
> Further evidence for this hypothesis comes from the dichotic listening experiments of Broadbent, in which simultaneously arriving auditory signals to the two ears are both decoded, though attention is directed to only one at a time." (pg. 295)

Our experience is not limited to what we have been directly conscious of. Consciousness is not a "force" that controls our behavior. It is simply an indication of which of our mental/neurological activity has the highest signal.

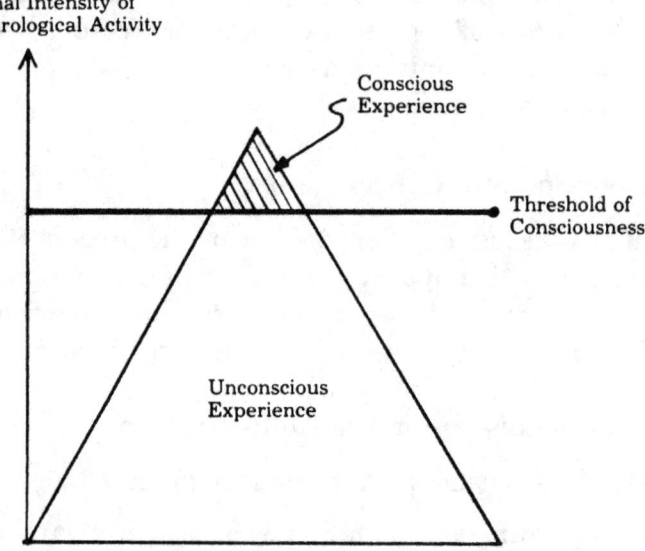

Figure 3.

In fact most of the limitations imposed by consciousness seem to be in an individual's ability to perform a conscious task rather than an individual's ability to process or analyze information nor on the actual amount of information an individual can consider at any point in time. Miller (1956) established that most people can consciously respond to about seven, plus or minus two, "chunks" of information in a single isolated trial, yet there is no established limit as to how much information can be stored in a single chunk.

Most tests on the limitations of memory and learning ability seem to be tests on *conscious* performance at particular points in time.

3. Experience about Experience

This is the level of experience that deals with the forms and patterns governing its own processes. At this level we are making models of the modelling process or *meta-models*. This type of experience involves the representation of how experience is represented: the structure of experience.

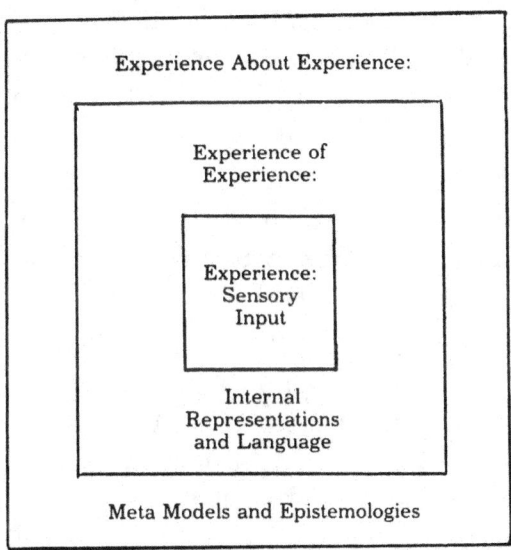

Figure 4. Forms of experience.

B. The Structure of Epistemology

The purpose of this discussion of experience has been to get some idea of the scope of a physiological and behavioral epistemology. From this discussion we may now derive the criteria for a satisfactory epistemology. Stated formally, an epistemology should provide for:

1. the enumeration of the entire locus of sensory

information that is pertinent to the understanding of behavior and experience—that is, the enumeration of all the *units of meaning* (the differences that make a difference) that will allow an individual to make *assumptions, deductions,* and *predictions,* and so on, about his/her own experience and that of other human beings.

2. the enumeration of *structural descriptions* (representations) of this sensory information. In other words, a delineation of representational hierarchies and levels, a phenomenology of objects, relationships, events, and so on.

3. the enumeration of the *causal circuits* underlying and creating the structural descriptions (the *formal* properties of representation).

4. a process model that is an explicit and systematic method of *applying* the patterns and principles claimed by the epistemology. An epistemology must not only be descriptive, it must be *useful* (applicable to one's ongoing experience).

5. a *feedback system*. It must not only be applicable but *accountable to an individual's experience* (it must maintain its proportionality).

1.2 Cybernetics

Cybernetic theory is essentially a meta-model (a model about modeling). Cybernetic models are different from statistical or linear models in that they deal with the feedback of total systems, systems in which events at any position in the system may be expected to have effect at all positions on the system at later times. In cybernetic models, a particular cause or effect cannot be isolated from its context. Therefore, each part must be considered and measured in terms of the whole. Human behavior and experience are undoubtedly the result of such a system. Therefore any satisfactory model of human experience, behavioral, physiological or epistemological, must be cybernetic.

What I would like to do in this section is present *an outline of cybernetic theory* as put forth by Gregory Bateson in his book *Steps to an Ecology of Mind* (1972).

A. Cybernetic Explanation

> Causal explanation is usually positive. We say that billiard ball B moved in such and such a direction because billiard ball A hit it at such and such an angle. In contrast to this, cybernetic explanation is always negative. We consider what alternatives possibilities could conceivably have occured and then ask why many of the alternatives were not followed, so the particular event was one of the few which could, in fact, occur. The classical example of this type of explanation is the theory of evolution under natural selection. . . .
>
> In cybernetic language, the course of events is said to be subject to *restraints*, and it is assumed that, apart from such restraints, the pathways of change would be governed only by equality of probability . . . the actual event in any sequence or aggregate is uniquely determined within the terms of the cybernetic explanation. Restraints of many different kinds may combine to generate this unique determination. . . . (pp. 399–400).

B. Units of Meaning

> . . . the subject matter of cybernetics is not events and objects but the *information* "carried" by events and objects. We consider the objects or events only as proposing facts, propositions, messages, percepts and the like. (p. 401)

1. Information and Difference

The elementary unit of meaning in perception is *difference*. Our sense organs (eyes, ears, bodies, noses and tongues) all "perceive" by responding to changes or differences in our environment. In order to see, for instance, our eyes are constantly making tiny microscopic movements. If they were ever to be held perfectly still, the cells in the retina would habituate to the unchanging stimulus and quit sending messages to the brain. In experiments where this has been done, subjects experience seeing the world literally fade away!

Physiologically, sensory perception takes place through *differences* in the location, firing pattern and interconnections of cortical neurons.

Behaviorally, if we want to discover what makes a good speller "good" at spelling, the most efficient way to proceed is to *contrast* the performance of the good speller with that of a poor speller. It is what the good speller does *differently* from the poor speller in his/her mind and with his/her body that will tell us the critical components of spelling. What the two do similarly is inconsequential with respect to spelling. This process of making comparisons, that we call *Contrastive Analysis*, is one of the primary modeling tools used in Neuro-Linguistic Programming.

The elementary unit of *meaning* in information, then, would be, as Bateson points out, "a *difference which makes a difference.*" And as Bateson goes on to point out:

> There are differences between differences. Every effective difference denotes a demarcation, a line of classification, and all classification is hierarchic. In other words, differences are themselves to be differentiated and classified . . . (p. 457)

2. Dimensionality and Chunking of Information

The basic dimensions of difference would be *space* and *time*. Spatial difference would involve the difference in location of two objects. Temporal difference would involve a change in that difference (your eyes moving with respect to the rest of your face, for example). A third type of difference would involve a change in the changing of the location of two objects (acceleration).

a. Classification of Difference

One can say that a classification of difference or information occurs at the point where a *different kind of difference makes a difference*. For example, speaking and writing are classified as different types of communication because one is dependent on differences in visual markers while the other carries information depending on differences in vocal utterances. In cybernetics there are two basic forms of classification: *hierarchic* and *logical typing*.

1) *Hierarchies* occur in the transition from microscopic to macroscopic information. The differences between *quantities* of information are often qualitatively different from the types of difference that create the *individual* bits of information. Thus as we move from *molecule* to *cell* to *tissue* to *organ* to *organism* to *society* we can see that the rules for the combination and interaction of the higher levels of the hierarchy are qualitatively different from those that govern the microscopic interactions. Other examples include:

    a) In mathematics, the numbers 8, 16 and 24 are a subclass of all numbers divisible by 4, which is in turn a subclass of all even numbers.

b) In behavior, an eye movement is a subclass of a facial expression, which in turn is a subclass of non-verbal communication, etc.

In hierarchic classification, however, one must keep in mind that each sub-unit is a part of the unit of next larger scope and that certain differences in the part of the unit have informational effect upon the larger unit, and vice versa.

2) *Logical typing* occurs where there is a *discontinuity* (as opposed to a continuity, as with the hierarchies) between levels of classification. This kind of discontinuity is exemplified:

a) in mathematics, by the restriction that a class cannot be a member of itself nor can one of the members *be* the class.

b) in logic, by the solution to the classic logical paradox, "This statement is false." (If the statement is true, it is false, and if it is false, then it is true, and so on.) The actual truth value of the statement is of a different logical type than the statement itself.

c) in behavior, by the fact that the reinforcement rules for exploration in animals is of a completely different nature than those for the process of testing that occurs in the act of exploration. As Bateson points out:

> . . . you can reinforce a rat (positively or negatively) when he investigates a particular strange object, and he will appropriately learn to approach or avoid it. But the very purpose of exploration is to get information about which objects should be approached or avoided. The discovery that a given object is dangerous is therefore a success in the business of getting information. The success will not discourage the rat from future exploration of other strange objects. (p. 282)

Beliefs and belief systems in human behavior also seem to have this type of structure.

3. Redundancy and the Coding of Information.[1]

Bateson writes:

> If from perception of some X, it is possible to make better than random guesses about some Y, there is "redundancy" between X and Y, or "X is a coded message about Y", or "Y is a transform of X", or "X is a transform of Y". Note that questions of intent or consciousness do not enter into these definitions . . . In general, semantic argument is minimized by isolating the variables in the drafting of the definitions. (p. 133, The CoEvolution Quarterly)

In *Steps to an Ecology of Mind* Bateson further elucidates the concept of "redundancy."

> The concept of redundancy is usually derived, as I have derived it, by first considering the maximum of information which might be carried by the given item and then considering how this total might be reduced by knowledge of the surrounding patterns of which the given items is a component part . . ." (p. 406)
>
> ". . . when an observer perceives only certain parts of a sequence or configuration of phenomena, he is in many cases able to guess, with better than random success, at the parts which he cannot immediately perceive (guessing that a tree will have roots, for example). It is, indeed, a principal goal of the scientist to elucidate these redundancies or patternings of the phenomenal world. (p. 414)

a. Coding

All coding is essentially of the part-for-whole nature of redundancy described above by Bateson: If we know some part of the information, we can, depend-

---

1. The majority of this discussion of coding is taken from Bateson's article, "Reality and Redundancy." Published in The CoEvolution Quarterly, Sum. 1975.

ing on our knowledge of that type of information, guess at the rest with better than random success. There are, however, a number of qualitatively different types of part-for-whole coding, and, as Bateson points out, "many characteristics of any information processing system are determined by the sorts of coding upon which that system depends."

1) *Causal and Correlative:* Here we can define the term "causation" as being synonomous to sequential correlation. Bateson points out that in the typical cause-effect explanation:

> We note *only* that when a given known "cause" is perceived, its known "effect" may be expected; and, vice versa, perceiving a known effect, we shall guess that its known cause preceded it. The commonest errors or pathologies of systems which rely on causal and correlative coding arise:
>
> a) from seeing only those causes and effects which are most obvious and ignoring the many other conditions which may be necessary to make the given "cause" effective, and ignoring the many other collateral effects which may be spread through the world in addition to the particular effect which the system "seeks;"
>
> b) from regarding the linkage between cause and effect as more direct than it commonly is; and
>
> c) from ignoring the circumstance that both "effect" and "cause" are usually (or always) components in *circuits* of causation. (p. 133, The CoEvolution Quarterly, S. 1975)

2) *Analogic:* This type of redundancy or coding depends on quantity and measurement (not upon numbers and counting). Some quantity in some X can be used to guess or predict some quantity in some Y. (p. 133, CEQ)

As is suggested by its name, analogic coding is where the part is a quantitative *analog* for the whole. An example of this type of coding would be the difference between "yes" and "yes!" The stress and volume with which the word "yes" is said is in most cases directly indicative of an individual's internal state of excitement. Similarly, the firing frequency of pressure receptors and neurons in a human being's somatosensory system increases in direct proportion to the increase in the amount of stimulus applied.

3) *Digital:* In this type of coding there is a discontinuity between the difference indicated by the part and the difference in the whole with which it is associated. Further, "those differences which make a difference are sharply defined contrasts with no intermediate values." In *Ecology of Mind* Bateson states:

> The signs themselves have no simple connection (e.g., correspondence of magnitude) with what they stand for. The numeral 5 is not bigger than the numeral 3. It is true that if we remove the crossbar from "7" we obtain the numeral "1;" but the crossbar does not in any sense, stand for the numeral "6." A name has only a purely conventional or arbitrary connection with the class named. (p. 373)

Examples of this discontinuity can be found in the difference between: a) the meaning and sound of "yes" and "no" b) the meaning and shape of the numbers "5" and "3" c) the meaning and the firing of "on/off" receptors in the visual and somatosensory (tactile) systems d) "0" and "1" in a binary system.

4. *Iconic:* This is the type of coding where the part of the code is a transform of a part of the information it is coding. A footprint in the sand indicates the presence of a human being. The frown on a person's face indicates the feelings in that person of the state of unhappiness of which the frown is a component part.

When, after an individual has been asked a question, he/she looks up and to the left and says, "Let me see," that sequence of events could mean: a) that the individual is actually "looking" (making internal visual images) for the information requested of him/her or b) the individual is communicating the information, "I am thinking about your question," by exhibiting a component of behavior that usually accompanies that person's internal process of thinking or remembering.

5) *Ostensive:* In ostensive coding, the part is a *qualitative* analog for the whole. For example, we can indicate the pitch of another person's voice by imitating that pitch with our own voice; we can lower our voice as we say, "The rock went deeper and deeper into the depths of the ocean." We can indicate where we want something put by pointing our finger in a certain direction.

Another example of ostensive coding is in the human nervous system where activity in the receptors in a certain location in an individual's body corresponds to activity of cortical cells in a certain location in that individual's brain.

6) *Evolutionary:* In this type of coding "information which is accumulated in organisms by evolutionary process . . . is complimentary to those environmental phenomena to which the organism must adjust." One interesting aspect of this type of coding is the hypothesis that much of the evolutionary development in predator-prey phenomena (animals which have developed bait organs or protective markings) is not simply the result of random mutation, but the result of some kind of genetic accumulation of information concerning a specific type of phenomena in the organism's environment.

A shorter-term example of this type of complimentary coding might be the rituals developed by some primitive tribes where the hunters put on the skins of their prey and assume the behavioral aspects of that animal. By assimilating the characteristics of the animal (by essentially becoming the prey for a time) the predator assimilates information that can be used in the hunting of that animal.

7) *Holographic:* This type of coding occurs in systems where there is a redundancy between all parts of the system; that is where the information carried by any part of the system affects the information carried by the other parts of the system. This type of system is called a *resonant system.* One example of this type of coding is the interference pattern or "beat frequency" created by the combination of two wave forms, that, itself, may be represented as a single wave form. All of the information about the two combined wave forms is contained in the interference pattern, and each of the two wave forms contains information about the resulting interference pattern. Information concerning either one or the other wave form may be attained by filtering out its companion from the interference pattern.

The moiré patterns that result from the overlap of two visual patterns are another example of this type of coding. In general, as Bateson points out, "When two repetitive systems are combined, a third is necessarily generated."

b. *Entropy*

Entropy is the opposite of information. It is characterized by the *total randomization of events* within a system: a situation in which there are no meaningful differences or redundancies. This type of system would be very rare, if possible at all, in that random response at one level will generate a non-random pattern of response at another level. As D. Layzer points out in his article *The Arrow of Time*, microscopic entropy at one point in space and time will affect (that is, make a difference in) the microscopic entropy at another point in space and time. This effect will make a difference at the macroscopic level.

So that, although a tossed penny will land with either heads or tails showing at random, if we consider this process from a different level we know that this series of random events will assume the pattern that the penny must land on either its head or tail side 50% of the time.

As Bateson states, insightfully:

> To the aesthetic eye, the form of a crab with one claw bigger than the other is not simply asymmetrical. It first proposes a rule of symmetry and then subtly denies the rule by proposing a more complex combination of rules. (*Ecology of Mind*, p. 410)

c. *Discontinuities and Catastrophies:* The Phenomena of Jumping Levels

Discontinuities, catastrophies, or thresholds are common phenomena in any system. Some examples are:

1. in neurophysiology, the "on/off" threshold properties of the firing of specialized receptors.

2. in atomic physics, the discontinuity in the jumping of electrons from one orbit to another.

3. in electronics, the threshold properties of semiconductors.

4. in behavior, when someone has the experience of "enlightenment" (on the positive side) or a "psychotic break" (on the negative side). Altered states of consciousness and amnesia may also be considered to be examples of behavioral discontinuities.

The nature of these events is generally summed up in the truism about the straw that broke the camel's back. In *Ecology of Mind*, Gregory Bateson gives further insight into this particular property of systems:

> All biological and evolving systems ... consist of complex cybernetic networks, and all such systems share certain formal characteristics. Each system contains subsystems which are potentially regenerative, i.e., which would go into exponential "runaway" if uncorrected ... The regenerative potentialities of such subsystems are typically kept in check by various sorts of governing loops to achieve "steady state" ... A constancy of some variables is maintained by changing other variables. ... Those mutational changes will be perpetuated which contribute to the constancy of that complex variable we call "survival." The same logic also applies to learning, social change, etc. The ongoing truth of certain descriptive propositions is maintained by altering other propositions.
> In systems containing many interconnected homeostatic loops, the changes brought about by external impact may slowly spread through the system ... Over time, the system becomes dependent upon the continued presence of that original external impact whose immediate effects were neutralized by the first order homeostasis ...
> In extreme cases, change will precipitate or permit some runaway or slippage along the potentially exponential curves of the underlying regenerative circuits. This may occur without total destruction of the system. The slippage along exponential curves will, of course, be limited, in extreme cases by breakdown of the system. (pp. 441–442)

C. Patterning

In conclusion to this discussion on cybernetics I would like to present a process involving the systematic application of the cybernetic principles. It is a system for working with systems that, I think, is particularly relevant to the integration of physiology and behavior.

In general the term "pattern" would be synonymous to the term "redundancy." As Bateson defines it:

> A pattern, in fact, is definable as an aggregate of events or objects which will permit in some degree (better than random) guesses when the entire aggregate is not available for inspection. (*Ecology of Mind,* p. 407.)

As a system, however, the process of patterning would consist of an enumeration of:

1. the basic units of difference and information pertinent to the functioning of the larger aggregate.

2. the set of rules governing the chunking and classification of this information.

3. knowledge of all the redundancy rules (types of coding of information) which are recognizable within the aggregate.

4. knowledge of the feedback properties of the aggregate and the aggregates of which it is formed and is a part.

It would be difficult, of course, if not impossible, to ever get a total enumeration of *all* the information listed above for any system at any one time, particularly that of human behavior. The reason being that no observer or group of observers will be able to discover or take into account all of the relevant information about the entire system governing the process of human behavior at any point in time—especially in a system that feeds back onto and changes itself. If, however, we consider that an observer might discover some of the relevant rules from his/her perception of less than the whole aggregate, "he," as Bateson proposes, "could then use his discovery in predicting rules for the remainder—rules which would be correct even though not exemplified." These predictions, of course, would be held accountable to the observer's own experience: past, ongoing, and future.

What I am proposing is a new way of thinking that involves the use of changing patterns dependent on the contextual conditions and feedback within and between systems existing at different points in time of an individual's ongoing experience. The patterns do not depend on statistical or linear cause-effect documentation, because (a) statistics leave out much important information regarding context, change, and feedback (patterns are not necessarily determined by quantities) and (b) documentation gives you no information about your immediate ongoing experience.

The process is, however, both formal and systematic and held rigidly accountable to the empirical evidence of one's own sensory experience.

1.3 Properties of a Cybernetic System

I wish to continue my presentation of the cybernetic model as follows:

a) Definition of the *basic units* of difference and the identification of those differences that make a difference.

b) Discussion of the restraints regarding the classification of information, that is, the *rules* regarding the chunking and formation of information into hierarchies and different logical types.

c) Discussion of coding restraints, that is, the rules regarding different *types* and *levels* of coding and redundancy and the types of information they communicate.

d) Discussion of *feedback* and interaction within and between sub-systems, possible causes of exponential slippage and other discontinuities, and development of new ways of coding in response to the elimination of others.

A. Units of Difference and Information

We know from physics that it is the difference in field potential between two points in time and space that constitute the essential units of structure of any physical phenomenon. It is how these differences are compared with other points in time and space that provides the measurement of information.

The basic carriers of this information can be identified as the sub-atomic particles (electrons, protons, and neutrons) which combine together, according to the properties of their respective charges, into the hierarchic structures and systems which compose the phenomenal universe; sub-atomic particles combine to form the different elements, molecules are formed from these elements; solids, liquids, or gases are from the molecules depending on environmental conditions and the structure and properties of the molecules. These structures then combine into larger systems and structures and so on.

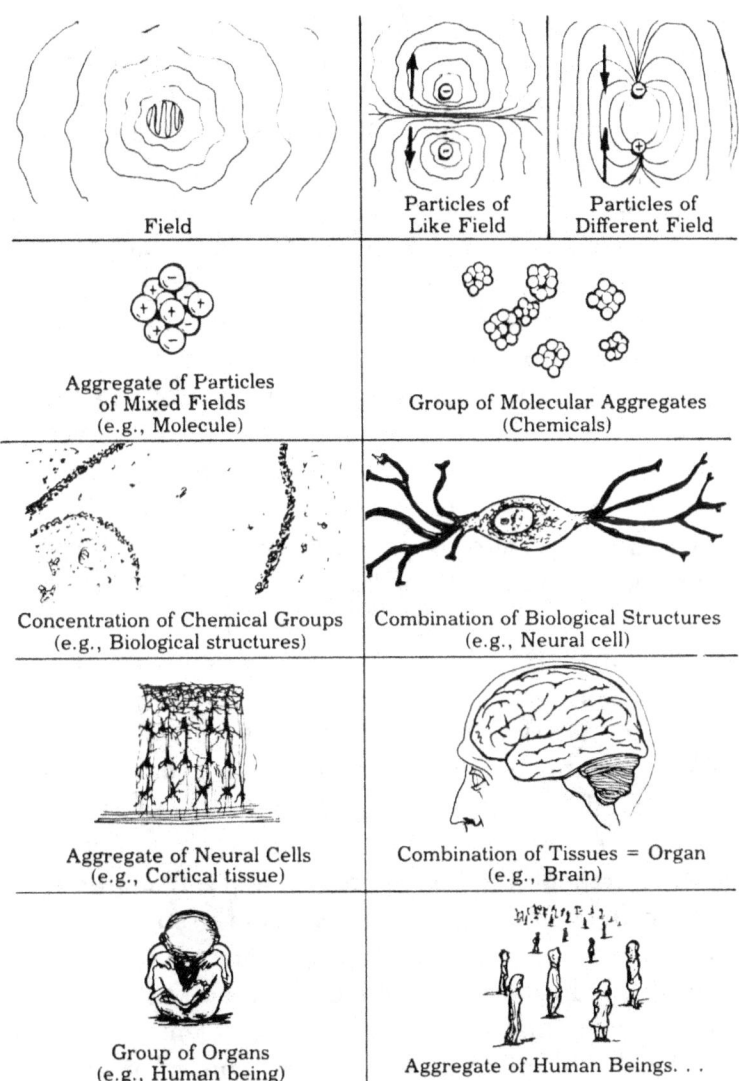

Figure 5. Hierarchical levels in human beings.

The point I wish to make here is that all of the chemical (ionic, hormonal, genetic, and so on) and biological (the structural and firing properties of neurons and groups of neurons, and so on) phenomena associated with the brain, are simply ways of rearranging the basic units of information (the fields associated with sub-atomic particles, particularly electrons) over time and space. The chunks or types of information one measures in the examination of the brain depends on which level of interaction one is observing. The total amount of information being received, analyzed, or communicated by the brain at any point in time is a function of the overall change in potential at the microscopic level within the volume contained by an individual's skull at any increment in time. Performance, or the facilitation of that information, would be a function of how those changes eventually affect the firing of motor efferent neurons (how the smaller changes in potential summate over time or space to produce certain larger order changes).

The implication of this is that the entire history of humankind is a function of the fluctuations of electrical fields within the volume contained by our skulls. Every human behavior from the construction of the Parthenon to the composition of a Bach fugue to rape and war to the construction of the space shuttle has been conceived of and implemented as a result of the displacement of electrical activity in the lump of gray matter that rests in between our ears. The evolution of human behavior, communication and intelligence occurs as we discover new and more effective ways to organize these fields of electricity.

B. Classification and Chunking of Information and Difference

Some of the coding restraints governing the classification and chunking of information have already been discussed in the previous sections. On the most basic physical level these restraints are the results of 1) the attraction of unlike charges, and 2) the repulsion of like charges. In larger-scale changes in potential these effects summate over time and space to produce either interference or resonance between different fields of potential. The amount of summation and subsequent interference or resonance is dependent on the size of the chunks (spatial or temporal) and the amount of information (difference in field potential in this case) or concentration of information carriers within the respective chunks or units being compared. As we begin to expand our parameters to the comparison of larger and larger "chunks," however, we must begin to take into account that it is practically impossible to totally isolate any difference or information within the confines of some particular spatial or temporal parameters. *Information is relative* in that (1) it is the *measurement* and comparison of difference (i.e., difference which makes a difference) with respect to some chunk of supposed neutral or null value or some value common to the respective differences being measured (ground); (2) depending on what one is measuring for, there will always be some *"noise"* inherent in the measurement; that is, (especially in the case of field potentials) the information in the chunk we are considering is going to be interacting with and changing with respect to the rest of the universe as well as the chunk we are measuring it against. Thus, we will call the amount of measurable difference between chunks "signal," and the external factors which limit that measurement "noise." The amount and accuracy of information measurement between chunks will depend on the respective *"signal-to-noise ratio"* (s/n) for those chunks. This ratio will vary in time as well as space.

In addition to the restraints dictated by the interaction between static charge over increments of time and space, there are restraints which result from the movement of charge *in* time and space i.e., electromagnetic field). An electromagnetic field is of a different logical type than an electrostatic field: it is a function of the *rate of change of position* in space of the information carrier with respect to time rather than a function of position of the information carrier with respect to other quantities of information carriers at any point in time. A difference that is the result of a time dependence as opposed to a difference that is purely spatial and identifies the most basic logical type distinction (See Appendix A).

Hierarchic classification involves identifying differences that occur as we compare (that is, find differences between) the information carried by larger and larger chunks, or units of space, in two or three dimensions or over larger intervals of time, or both. We will call information that is created or measured as a function of this type of classification information of a different *logical level.* Information of a different *logical type* will be understood as differences that exist regardless of the quantitative aspects of the chunks being compared.

As I pointed out earlier in this paper, in hierarchic classification each sub-unit will have an informational effect on the unit at the next level up and vice versa. Information of different logical types also has informational effect on one another, although not of the part-whole relationship that exemplifies the interaction between levels. The informational effects between levels and types is called *feedback* and is probably the major distinguishing feature of cybernetic systems. The remainder of this section will consist of an elaboration of (a) the dimensions and parameters distinguishing the informational content of logical types and logical levels and (b) the interactions and feedback properties between logical types and levels.

1. Dimensionality and Measurement

My goal in this subsection is to present and establish dimensions that are formal enough to be isomorphic at all levels of interaction. In other words, formal enough to apply, no matter how big the chunks one is comparing. We have already, in this paper, identified and discussed what I believe to be the three basic dimensions for the classification of informational elements of distinct logical types:

(1) *Spatial relations:* These will be defined as differences resulting from the comparison of two or three dimensional areas within some static increment in time, or areas that retain a relatively static relationship over some interval of time.

(2) *Temporal relations:* These are differences that occur as the result of the comparison of consecutive increments of time. Although these differences may be represented or identified by spatial relations, they must occur as a *function* of a fourth dimension.

(3) *Causal relations:* These are defined as differences which result from the co-occurence of the other two relations, that is, differences due to the proximity and sequencing of information carriers with respect to each other.

a. Descriptive parameters

Keeping within these relations, we can now move to the consideration of the measurement of the respective differences brought about by these relations:

(a) *Signal-to-noise ratio* (S/N): as was pointed out earlier, this involves the comparative amplitudes or intensities of some difference of informational value relative to some observational point and some other difference which co-occurs and interferes with that information. Note that the ratio may vary with temporal relations as well as with spatial or causal relations. Its importance to the neurological model being developed here can be seen in its function as an indicator of degree of excitatory and inhibitory interaction, respective density of cortical cells in brain regions designated to perform modality-specific functions, the degree of dominance certain inputs will have in determining behavior, acuity of distinctions an organism can make about its environment, and so forth.

(b) *Simultaneous or sequential expression:* This parameter measures information and difference that is time dependent. Its importance can be seen as an indicator/determiner of various S/N ratios and the effects thereof. For example, depending on whether two inputs are received simultaneously or sequentially, the S/N of their representations with respect to each other may vary significantly, resulting in a competition for response dominance, or perhaps a representational integration rather than a separation. Similarly, parallel processing, as opposed to sequential processing, may result in differentiation or lateralization of function.

(c) *Afferent versus Efferent:* This parameter traces causal chains and distinguishes between *input* and *output* interactions. It implies, of course, some sort of transformation function, the nature of which will be discussed in more detail later on. This parameter measures the relationship between stimulus-response phenomena, the structural activity of neural aggregates, competence/performance differences and to some extent internal versus external locus of behavior.

Barrett (1968) provides a good example of the use of these dimensions and parameters in his comparison of the cortex of the brain to an interferometer (a device used to make precise measurements in minute changes in the wavelengths and waveshapes of light). Pointing out that electrical input to the nervous system can be exhaustively described at any point in time in terms of amplitude, frequency (oscillation in amplitude over time), and phase (position of frequency in time) relationships, he claims that optical information can be interpreted and stored in terms of amplitude contrast (S/N) and phase contrast (simultaneous/sequential) and their actualization in the central nervous system (afferent-transform-efferent relations). This work will play an important role later in this paper when information "coding" is discussed.

2. *Classification and Interaction*

Before discussing the properties of interaction between information of different classifications (both levels and types), some further qualifications and distinctions are in order:

(1) As was declared earlier, all measurement is relative to some *reference structure*. The emergence of relativistic mechanics over classical Newtonian physics provided in itself a new reference structure by which to measure and interpret the entire locus of physical phenomena. Therefore, the classification of differences in information will itself be relative to some reference structure. Further, as the information being chunked becomes more and more complex, changes in only one parameter may create the difference that makes a difference with respect to some particular reference structure. For example, one may not consider two different frequencies of electromagnetic waves to constitute a difference in logical types, rather, they have a hierarchic relationship with respect to the amount of oscillation of amplitude values per second. Yet, if one of the two waves represented a point on the visible color spectrum and the other an invisible radio carrier wave, we may wish to define them on the perceptual level as different logical types.

Another name for reference structure is *context*.

(2) It was claimed earlier that *logical type* differences remained invariant across levels. The electrical, chemical, and biological interactions that take place in the brain can be said to be of different logical types each with its own particular dynamics of interaction and structural properties which remain distinct at all levels of neural function and, indeed, in all biological systems. Yet the differences that distinguish them as logical types do not appear until a cer-

tain level of chunking information, but remain invariant after that.

(3) A change in logical type can occur by what was called a *"catastrophe"* earlier in this paper. The change in certain forms of matter from a solid form to liquid form and to gaseous form at certain energy levels (boiling point catastrophe, melting point, and so forth) are an example of this.

(4) A *logical level* distinction will be redefined as being necessarily *self-referential* as well as occurring as a function of the size of a chunk, as opposed to a strictly *hierarchical level*. Learning about learning, talking about talking, and so on, would be examples of information at different logical levels where a certain process can assume a *meta position* with respect to itself.

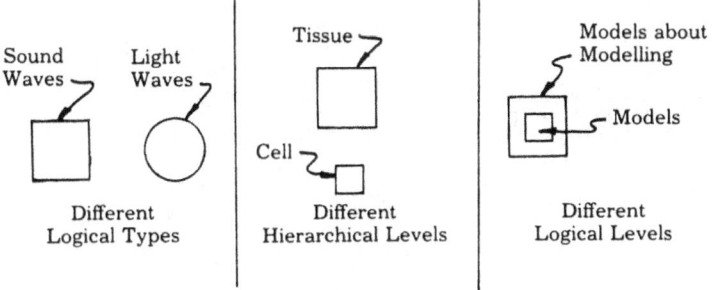

Figure 6.

a. Properties of interaction

In this section I will discuss what happens when information of different or of the same logical types and levels interact. The form of discussion will consist of (1) presentation of the pattern and (2) examples of the pattern from (a) sub-atomic systems

(b) systems of electronic components (c) neurophysiology (d) systems of behavior (e) systems of language and logic and (f) mathematical relationships.

(a) *Differences of the same logical type interacting with each other at the same hierarchic or logical level* produces direct *interference* (positively or negatively). When difference pattern A and difference pattern B are combined in the same time-space coordinates, they each alter the shape frequency and amplitude of the other in an additive or subtractive way to produce a third pattern at the same logical level, although possibly of a different logical type, depending on one's reference value. (See figure 7.)

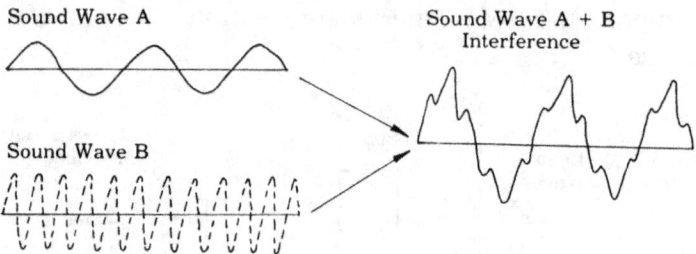

Figure 7.

1) Two particles of like field (electrons, for example) will produce a third field composed of the addition or subtraction of their fields depending on whether they are moving in similar or opposite directions. If they are moving in the same direction, their fields will add; if opposite, their fields will subtract.

2) Two electric signals in the same circuit, two sounds uttered in the same room, two visual images superimposed on each other will exhibit the same

Roots of Neuro-Linguistic Programming

pattern of constructive or destructive interference.

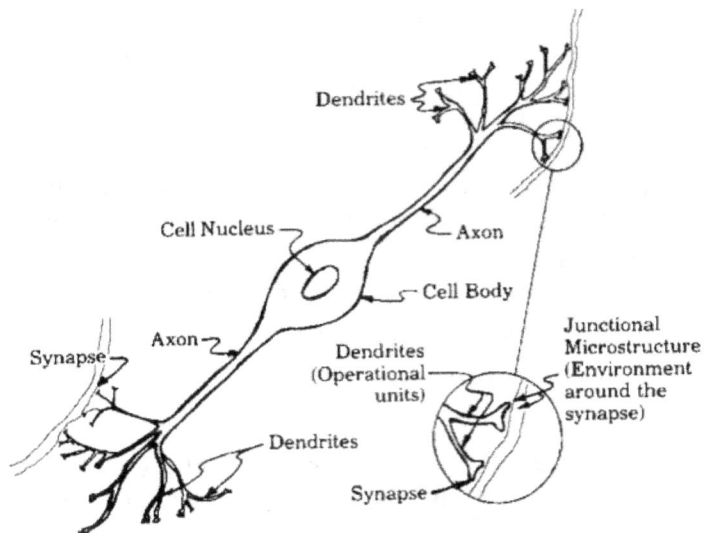

Figure 8. The Neuron: Basic carrier of information in the brain.

3) Karl Pribram (1971) defines two different logical types of information carriers in neural systems: (1) *individual operational units* (that is, individual neurons of specified membrane potentials and dendritic and axonal interconnections) and (2) the *junctional microstructure* (the chemical and electrical sensitivity of the environment surrounding the cell and its synaptic junctions). Individual neurons, depending on the excitatory or inhibitory nature of their interconnections, will fire in patterns that will summate spatially and temporally on a third neuron that will fire as a function of the total excitatory less inhibitory activity of synaptic input from other neurons. Similarly, the neural environment will vary as a function of the summation of the var-

ious concentration gradients of the ions that compose it. As we shall see later, however, operational units do not interact with their chemical environment in the same additive fashion.

4) In logic we may call the interaction between statements of the same logical type on the same logical level either *contradictory* or *supportive* or *equivalent*. If we say $A = B = C$, then $A = C$. However if $A \neq B = C$, then $A \neq C$. A proposition cannot equal its negation: $P \neq \sim P$. For example, the statement "This sentence is true and it is not true," is contradictory.

5) In behavioral terms we may define the interaction of two behaviors of the same logical type at the same level to be *symmetrical* with respect to each other: When one animal exhibits a behavior of class A the other exhibits a behavior also of class A—one person yells and the other yells back, for instance. Depending on the contextual variables the behaviors will coincide constructively or destructively.

6) The mathematical correlate of the proposed pattern of interaction would be the relationship of integers on a number line or any linear function on a two-dimensional relationship. The interaction of A and B results in $A + B$ or $A - B$.

(b) *Differences of different logical types interacting on the same logical level produce coexistence of functional homogeneity.* (See figure 9.)

By Bertrand Russell's definition:

> A *type* is defined as the range of significance of a propositional function, i.e., as the collection of arguments for which the said function has values.

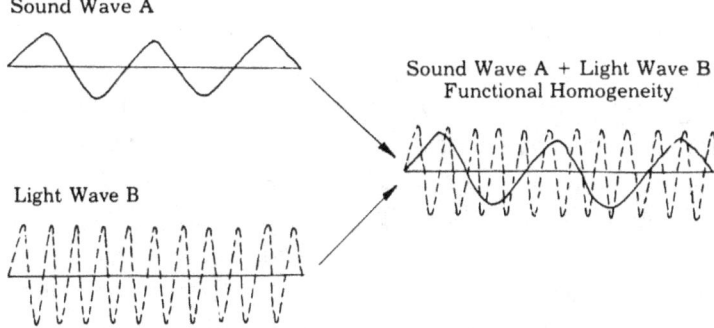

Figure 9.

In the terminology of this paper, a "propositional function" is the proposed difference that makes a difference. And, by Russell's definition, it should be fairly obvious that if the functional significance of two types of information is different that information should be able to coexist in the same time-space coordinates with no interference of informational significance. Both types will retain a logical homogeneity. Some examples of this include:

1) All elementary particles have properties of (carry information about) both wave and particle values.

2) Because an electromagnetic field is generated as a function of the time rate of change of electrical activity (i.e., current), any electrical activity contains some value of both static electric field and electromagnetic field simultaneously.

3) A individual neural receptor can contain and transmit information concerning location of stimulus, intensity of stimulus, location on the organism's body that the stimulus is being sensed, type of stimulus, and so on.

4) Russell, of course, used this distinction to dispel what appeared to be logical paradoxes by establishing first and second order propositions as different logical types. He claimed, "Whatever contains an apparent variable must not be a possible value of that variable." Thus, when Epimenides claims that he is a liar his statement need not necessarily apply to itself. Instead, he asserts, "all first order propositions affirmed by me are false." He has asserted a second order proposition and thus no contradiction arises. Both types of information occur at the same time.

5) Complex forms of human behavior, such as play or metaphor, can carry a number of different types of information. Greek myths, for instance, carry historical information about Greek culture but simultaneously transmit and install thinking strategies (i.e., teach a lesson that will affect ongoing behavior).

6) The most common example in mathematics is the rule that a class cannot be a member of itself. The most basic logical type distinction would of course be that between static and dynamic aspects of a proposition or equation: that between the term or *subject* of the proposition and the concept or *predicate* (operator).

Interaction of different logical types should not be mistaken for *confusion of logical types* at the same level, which results in a conceptual *paradox* or *oscillation*. Confusion of logical types usually occurs when there is a possible overlap of functional domains in the representation of the relationship. For example, the statement " 'Cat' is a four-legged animal and a three-letter word," causes confusion of logical types because the digital representation "cat" may be confused to be the actual referential index of both the predications. In actuality the four-legged animal refers to the domain of representation which the word "cat" indicates.

Confusion of functional domain becomes simple enough when enough information is deleted from the representation. The statement "this sentence is false" contains no information concerning the reference structure by which its "truth" can be judged or how.

Lamendella (1977) in his criticism of the locationalists' search for memory centers or behavior or learning centers provides a good example of how logical types have been confused in the study of neurophysiology in the past. In this case, perceptual modalities, which can be shown to display higher signal-to-noise ratio in certain areas of the human cortex, were confused to be of the same functional domain as other brain functions such as learning and memory, which in reality involve spatial and temporal patterns *between* perceptual modalities. The result was a search for learning or memory storage centers, rather than the means and rules for perceptual interaction.

c) *Differences of the same or different logical type interacting at different levels (hierarchical or logical respectively)* will result in the *modulation* of the difference on the lower level.(See figure 10.)

In this case, difference pattern A (which may or may not be of the same logical type as B depending on reference structure) is *meta* to B, modulates B, in that it systematically changes one parameter of B, altering the amount of information B is carrying rather than changing it, as in the case of interference. Note also that the information content of A exists entirely intact, but is of a different logical type than B.

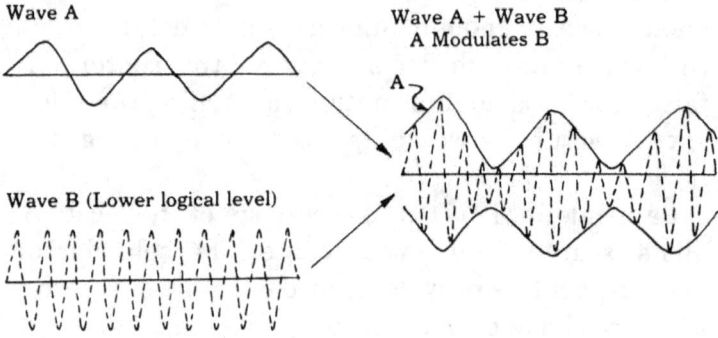

Figure 10.

The creation and classification of levels can be defined as the result of applying Russell's process of generalization to some proposed difference which makes a difference. Russell defines "generalization" as:

> ... the substitution of a variable for one of the terms of a proposition, and the assertion of the resulting function for all possible values of the variable. A proposition containing no apparent variables we will call an *elementary* proposition ... all generalized propositions presuppose elementary propositions.

By making the difference between two variables a variable one creates another level of difference. Levels can be created statically, by purely structural relations, or dynamically by the interaction of structures over time. As was noted earlier, and will be discussed again later, a change in levels may or may not constitute a change in logical type, depending on context.

Examples of levels include:

1) Early particle physics was based on the interactions of a number of sub-atomic particles with differences of primary logical types (for example, electrons, protons, neutrons, photons). Different types of atoms were created by using these particles as variables and combining them in different ways. Molecules, the next level, were constructed with atoms as variables. Chemical and biological phenomena had molecules as variables. The lower level structures are modulated in that they retain their logical integrity, yet their spatial, temporal, and causal interactions with other particles are significantly altered in the presence of higher level structures.

2) Transmission of visual and auditory information in television is accomplished by varying some electric signal as a function of another. Amplitude and frequency modulation are possible when one signal is set as the variable of another. The variable in this case is called a *carrier*. An interesting point to note is that any signal can be placed meta to another. This property is essential for decoding of information.

3) E. Roy John (1975) constructs a series of levels of neural information processing in *A Model of Consciousness* where "invariances across the representation of multiple items on the same level, (which share a common informational feature) constitute the representation of information on a higher level."

> 1. *Sensations* are the spatio-temporal patterns of information arriving in the central nervous system because of excitation of exteroreceptive and interoceptive organs. They are a product of the irritability of living matter and constitute first

order information . . . sensations can elicit reflex responses adjusting the organism to its environment.

2. *Perceptions* are the interpretation of the meaning of sensations in the context of stored information about previous experiences. Perceptions constitute second order information resulting from the interaction between sensations and memories . . . (perception involves) adjustments of the organism to its environment as a function of the experimental context of a stimulus rather than the action of the stimulus alone. (p. 4)

Perceptions are an operator on primary sensory experience which affect behavior in terms of relational schemata between the variables of direct sensory stimulation.

John defines a third level which he calls *consciousness* where perceptions become the variables and form a "unified multidimensional representation of the state of the system and environment, integrated with stored information generating emotional reactions and programs of behavior for adjustment." It is at this level that propositional activity begins to take place, particularly those involving causal relationships and events. It is at this level that John claims attention is focused, perceptions influenced, emotions aroused, drive priorities altered, and plans of behavior altered.

The level in which conscious experience and propositions become variables, John calls *self* and self-awareness. He defines "self" as the level in which previous perceptions and conscious programs may be arbitrarily organized rather than occurring according to a previously established sequence.

4) In logic, one might say that modulation corresponds to *qualification* The higher level statement,

such as "I am lying" qualifies the statements of a lower level such as "my name is George," "I am 38 years old," "I live on Mars," etc.

5) In behavioral interactions we may say that when the behavior, A, of one organism limits and controls the behavior of another, B, even though the two are functionally distinct, then A is of a higher level than B. This type of relationship can also be called *complimentary*. Examples of this would include a rider directing his/her horse by pulling on its reins; an employer verbally directing an employee or a conductor directing an orchestra.

6) Mathematically, such relationships would most likely be represented in terms of exponential or trigonometric functions, where B varies in some power relationship under A (that is B^A).

b. Feedback

Feedback properties are extremely important to relational schemata especially where matters of functional significance are involved. Feedback properties identify the boundaries of a system and account for what might be called the functional overlap between types and levels.

The most influential property of feedback will be whether it is a function of the same or different levels, with a special case occurring where functions at different levels are self-referential. A system where feedback occurs only at one level is called a *closed* system; a system where feedback can occur between levels is an *open* system.

(a) *Feedback between similar logical types on the same level* would most likely result in a *resonance or dampening* with the input pattern. This type of feedback occurs where system is a causal chain in which, via some link in the chain, the informational impulse (or some transformation of it) is fed directly back on the initiator, with either positive or negative interference. Two people pulling on each end of a two-handle saw, for example, will either help or hinder one another in sawing a log depending on how much they are able to synchronize their feedback. Similarly, groups of singers, dancers or musicians will create a resonance or interference depending on their abilities to keep in phase with one another. Many electrical and neural circuits are probably constructed this way. Another result of this interaction might be *homeostasis* or some kind of variation around a homeostatic base line.

(b) *Feedback between different logical types on the same level* will result in *oscillation*. Many people experience oscillation of this sort when two different logical types of information (two sensory systems, for instance) are applied to the same subject during a decision making strategy. On the one hand, something may *"look"* good, but at the same time it just doesn't *"feel"* right. A capacitor and an inductor in series in a closed circuit is probably the best paradigm for this type of interaction: The capacitor (which stores electrostatic energy through the interactions of opposite fields on two separate conductive plates) discharges across an inductor (a coil of wire that generates a coherent electromagnetic field in proportion to current flow) when a switch is closed. When the two storage plates have reached equilibrium, there is no more current flow, the magnetic field collapses (reversing its direction), facilitating a cur-

Roots of Neuro-Linguistic Programming

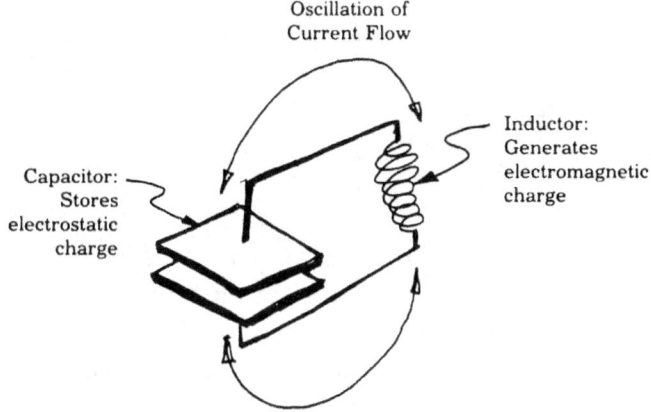

Figure 11. Feedback loop between different logical types of electric fields.

rent flow in the opposite direction also, recharging the conductive plates, and the interaction begins again.

Oscillation also occurs between informational elements of different types in neural systems when a neuron fires. When an "action potential" is initiated in an individual neuron there is an oscillation of electric potential as ions from the cell's environment rush in and out of the cell boundaries as a function of changing chemical and electric concentration gradients and fluctuations in membrane permeability.

(c) *Feedback between different logical types on different levels* results in *reciprocal modulation* of participating elements.

The phenomenon of the "self-fulfilling prophecy" is a result of this kind of feedback. The belief that someone is "stupid" or "intelligent" is of a different logical type than that person's actual behavior—the belief is a mental activity while the behavior is physical. The two should have no functional effect on one another. Believing that a chair can dance should not necessarily cause it to dance. Believing that someone is stupid or smart should not affect their intelligence. However, when a feedback system is provided the two may begin to shape each other. A dramatic demonstration of this was made in an experiment where a group of children who made average scores on an intelligence test was divided in half. One half was assigned to a teacher who was told that the children were "gifted." The other half was assigned to a teacher who was told that the children were "slow." At the end of the year, both groups were retested. The group that was given to the teacher who believed they were "gifted" scored above average. The group that was believed to be "slow" scored below average. The beliefs that the teachers had modulated their behavior. This in turn created responses in the students which reinforced the belief and so on.

It should be noted that because the belief is of a different logical type than the specific behaviors it is modulating, a counterexample to the belief will not necessarily interfere with it or challenge it. Thus, if a student who is believed to be "slow" scores well it might well be reframed as a function of "luck" or "cheating" and not challenge the basic premise of the belief.

The interaction between a neuron and its junctional microstructure as a function of slow-wave potentials is a good example of this relationship, for those readers who are familiar with neurological terminology. The slow-wave potential (an electrical brain rhythm presumably generated by lower brain structures) affects the sensitivity of the junctional microstructure between the synapses (point of connection) of cortical neurons by modulating the strength of the field gradient in the ionic environment. These fluctuations in sensitivity will either positively or negatively reinforce the transmission of a nerve impulse across the synapse. When and if the neuron fires, however, the resulting displacement of ions in the environment will cause local changes in field gradient also affecting the sensitivity of junctional microstructures and the probability that the cell will fire. So that the neuron's environment modulates the firing of the cell, and the firing of the neural cell modulates the properties of its environment.

(d) *Feedback between similar logical types at different logical levels* results in a *metacomplimentary relationship*. This type of relationship, when we apply Russell's definition of generalization to relations with some kind of functional overlap, can be considered to be one in which the relationship itself is one of its own variables. The metacomplimentary relationship occurs because, though the higher level function is modulating the variables at its subordinate level, it is itself being modulated as a lower level function. An interesting example of this is when one individual tells another to control him. By directing the other person to control him, the individual is guiding or controlling the behavior of the other on a metalevel. In a representative government, for instance, a group of people (the citizens) are given the ability, to a certain degree, to govern and control the group of people who govern and control them through the feedback system created by the voting process.

Self-reflexive levels are particularly common in representational hierarchies: making models of models or models about models, representations about representations, talking about talking, thinking about thinking, and so on.

A good example of functional overlap causing a metacomplimentary relationship is talking *about* a relationship. Talking about something is generally considered of a different logical type than the thing itself (for example, talking about a cat biting you cannot physically hurt you). Talking about the relationship you have with another individual, however, may be subsumed under the functional significance of that relationship. So even though talking about a relationship may be meta to that relationship it may itself be modulated by the interactive rules defined by the relationship.

C. Coding of Information

Coding of information involves the transformation and representation of differences organized and classified according to the dimensional parameters, types, and levels discussed in the previous section. The transformation and interaction of information in the human nervous system takes place statically in terms of spatial and structural constraints at particular points in time, and dynamically in the interaction between structures over time. In the next section, I will first address the structural aspects of coding in neural systems and then proceed to analyze the dynamic implications of coding that these structural constraints produce.

II. Towards a Cybernetic Theory of Brain Function

In *Physiological Psychology* (1970), P. M. Milner states that:

... if one *theory* of neural function should prove wrong and be replaced by a better one, serious changes in psychological thinking do not follow. The discovery of new neural *phenomena*, on the other hand, may entail profound changes in the thinking of physiological psychologists, opening up new possibilities for model building. The most important discovery that we can look forward to is one that would provide the basis for an explanation of learning, the storage of information. (p. 19)

In this section of this paper I would like to present both new neural phenomena and new theory concerning neural functioning and discuss the implications these have toward the building a new model of psychology and brain function:

1) The discussion of new *phenomena* will be centered around the statistical brain theory proposed by E. Roy John that claims that no one mental function can be attributed to activity in any specific cell or group of cells, but that vast regions of the brain are involved in any thought process with some regions contributing more than others to any given function.

2) The model will be a cybernetic model of brain function, combining the new information to be presented with the other information available about brain structure into a total cybernetic system, that is, a system composed of and affected by a hierarchy of sub-systems and which is itself a part of a hierarchy of larger systems. This model will be used to discuss and explain some of the phenomena of learning and other behavioral structures and events such as the development within individuals of hemispheric specialization, and the process of *"anchoring"* (the association of one experience with another).

2.1 Some Previous Theories of Brain Function

Most contemporary theories of brain function equate specific physical and mental activities with specific areas in the brain: certain groups of cortical cells are associated with specific sensory receptors or motor efferents. Any perceptual analysis, learning or the accumulation or communication of information, then, can only occur by some kind of connection between these groups of cells. There have been a number of theories purported as to how this takes place:

A. Generation of New Connections

These theories maintain that when something is learned, new connections appear between nerve cells in different parts of the brain. Remembering or carrying out some learned response is the result of the reactivation of the connected cells. There seem to be two basic hypotheses about how these cells become connected:

1. One claims that the new connections are actually made by the growth of new nerve fibers between cells as sequences of actions or perceptions are learned or remembered.

2. The other claims that synaptic activity between two neurons is more readily generated if there is consistent activity between the two. Learning or memory, then, would be a function of how much and how often these cells are fired in sequence together. This type of hypothesis would imply a degeneration or even loss of memory or learned response with re-

spect to time or relative inactivity of these particular connections.

B. Cell Assemblies

Milner (1970) points out that, in higher animals, the motor system has access to sensory input from all modalities that appear to be pooled at some level, at which each has the power to elicit a common neural activity representing the object. He cites Hebb (1949, 1966) for the explanation:

> In his treatment of perception he (Hebb) speculated that sensory input from a class of objects might establish a pattern of cortical firing (he called it a *cell assembly*) that would eventually become semi-independent of the input originally responsible for building it and would become capable of firing in response to other modes of input from the same or similar objects. In other words, a particular group of cortical neurons might first be excited by visual stimuli from triangles, but many of the same group would fire later if triangular objects were touched, three dots were seen, and so on . . .

It would seem, however, that in this kind of explanation, as in the two above, that (a) there would be considerable time required for the consolidation of memories, concepts, or other types of learning, (b) these connections would be subject to obliteration by lesions or ablasions within or between different areas of the cortex, and (c) they would place considerable limitations on human beings' abilities to change conceptual or perceptual associations or recover from brain damage or other cerebral traumas.

These kinds of conclusions do not fit with much of the research done on lesions, ablasions or other types of brain damage, nor with my own experiences with hypnosis and Neuro-Linguistic Programming.

E. Roy John points out:

> Some researchers have erased memory by cutting parts of animals' brains. But a careful look at these experiments usually shows that the brain damage caused less specific deficits—changes in motivation, attention, or sensory sensitivity—and not the loss of specific memory.
>
> Also, a great deal of research has shown that when brain damage that would normally cause a loss of certain capacities is inflicted in stages there is little or no loss. For example, if both sides of the visual system of a rat's brain are destroyed in the same operation, the rat loses the ability to recognize visual patterns. But if the second side is destroyed two weeks after the first, the rat can still recognize patterns. Apparently, during the two week interval, the rat learns to use other parts of the brain for recognition . . . . (Psychology Today, May 1976, p. 48)

John points out that the localization theory provides no explanation for this phenomenon nor for the fact that people with severe brain damage can sometimes recover lost functions completely, since destroyed brain cells and nerve tissues do not regenerate.

C. Causal Loops

Milner attempts to account for these types of phenomena with a theory of causal loops.

> A neuron at point A in the cortex, let us say, sends an axonal branch to point B, where it synapses with several thousand other cortical neurons. Some of the latter will be fired by the first neuron, and let us suppose that they send axons to points C, D, E, and so on, where each synapses with thousands of new cells. Sooner or later, some cells that send branches back into the region around A and synapse with the first cell are bound to be reached by the activity. Other neurons may go to points B, C, D, and so on and form loops within the larger loops. (*Physiological Psychology*, p. 117)

Milner goes on to maintain:

> ... once a small fraction of neurons have been recruited to snowballing activity of this sort, they drive up the level of inhibitory activity, which prevents further spread; a controlled burst of firing is thus confined to a limited group of neurons. A short time later the activity dies down (because of either fatigue, adaptation of the synapses, or inhibition from cell beginning to escape from the inhibition set up by the group) and a similar burst of activity establishes itself in another group of cells. ... there is an endless sequence of brief bursts of firing in one complex loop after another in the cortex, each involving thousands or perhaps tens of thousands of neurons. These networks would fire due to their interconnections whether or not there was input to the cortex, but, of course, input, when present, would have an important effect upon the pattern of activity. (p. 117)

Milner postulates that these networks would eventually become distributed throughout cortex and become accessible to inputs and different patterns of input from many modalities, and that, because of the multiplicity of internal connections and loops, they could not be eliminated by small lesions in specific areas of the cortex. He claims, "such a lesion would destroy perhaps 10–20 per cent of the network, but in a well established assembly the remaining connections would be enough to carry on."

Although this explanation is certainly admirable, I do not believe that it accounts for many of higher levels of learning or experience, nor many of the equipotential aspects of the cortex.

D. Digital Circuitry

Another interesting, though incomplete, theory of neural function is one that compares brain function and organization to the circuitry of computers. Making an analogy between the threshold properties of individual neurons, the phenomena of spatial and temporal summation, and synaptic inhibition to the operation of gates in the nand, and inverter circuits in computer chips, these theories compare the types of information and information processing carried out by neurons, to those carried out by the binary circuitry in computers.

This type of theory, however, necessarily implies a mechanism for the storage of memory that could be played over just as magnetic tapes are used in computer technology.

This theory and all the other theories presented thus far in this section seem to fail to account for a great deal of mental and neural phenomena, some of which will be presented in the next section.

2.2 The Statistical Brain

The statistical brain theory, as I pointed out earlier, was formulated by E. Roy John.[2] He writes:

> The earliest electrophysiological indications that memories do not consist of new connections between cells came when Keith Killam and I implanted 34 electrodes in different parts of the brain of a cat. These electrodes let us see the electrical rhythms recorded from many parts of the brain as the cat watched a light that flashed repeatedly. Some of the regions showed electrical waves at the same frequency as the flashing light. We called these waves "labeled rhythms." These labeled rhythms were the response of the cat's brain to the flashing light; other regions showed only random activity.
>
> When the cat first saw the flashing light, labeled rhythms showed up in only a few parts of the brain, mainly in areas included in the classical anatomical sketch of the visual system. But as the cat learned to associate the lights with the

2. This discussion of statistical brain theory is taken from an article by E. Roy John in *Psychology Today*, May 1976, pp 48–52.

need to jump a hurdle to avoid shock, the lights took on new meaning. The labeled rhythms spread to other parts of the brain . . . (p. 51)

The parts of the brain to which the labeled rhythms spread were reported by John (1964) to include all the sensory cortex, association cortex, sensory relay nuclei, the thalamic and mesencephalic reticular formation, and rhinencephalic structures. (See Appendix A for a discussion of how electrical field effects may contribute to the spread of the labeled rhythm.)

John also reports that once the labeled rhythms spread to other parts of the cat's brain, they would sometimes appear in absence of the sensory stimulus, and that sometimes the cat accompanied these bursts of labeled rhythms with the learned action that had become associated with them.

By calculating the average evoked potential (a computerized measure which calculates the average electrical response pattern from a certain brain region and represents it as a specifically shaped wave) John and Killam were able to depict many of the cat's learned responses as specifically shaped wave forms, and to tell which part of the wave shape came from the brain receiving sensory information and which from memory. John reports his results:

> We found a lawful relationship between the senses and memory in each region: the more a brain region senses an event, the more it remembers it.
> We also found that brain regions do not make an all-or-nothing contribution to mental operations. Each brain region makes an average contribution to practically every operation, but these contributions are graded, each region more involved on the average in some operations than others. (p. 52, P.T.)

Using microelectrodes to measure the response of individual cells to external stimulus, John found that every cell recorded "responded in unpredictable ways to new stimulus, more predictable ways to familiar stimuli, and in different ways to different stimuli." John also found that cells showed much random activity, and, more importantly, that well-separated cells in distant parts of the brain sometimes responded to the same stimulus with the identical firing rhythms. Further, the firing pattern reflected what the cat would later do and not the stimulus being given it.

The finding that cells in different parts of the cortex fire in response to the same stimulus seems to fit with the results of Bach-y-Rita's research. In his book *Brain Mechanism in Sensory Substitution* (1965), he states:

> Indeed visual responses have been reported to appear earlier in the somesthetic cortex than in the specific visual cortex (Kreindler, Crignel, Stoica, and Sotirescu, 1963). Similarly, responses to stimulation of the skin can be recorded from widely varying regions of the cortex, including "specific" somatosensory cortex, association areas, and even the visual cortex (Murata, Cramer, and Bach-y-Rita, 1965).
>
> In a study of the cat's primary visual cortical cells, Murata, et al., (1965) demonstrated that even these cells were polysensory, with approximately 37% of them responding to auditory and 46% to skin stimulation, compared to the 70% responding to the visual stimulus we employed. Most of the units responding to auditory and visual stimulation also responded to the skin stimulation . . . . These results demonstrated that the visual cortex (the cortex considered most highly specialized of the sensory projection areas) received input from other sensory modalities as well as visual input, and this suggests an associative or integrating role of at least some of the cells in this area.

Roots of Neuro-Linguistic Programming

It is also interesting to note John's finding that much of the firing of cortical cells was not in response to direct sensory stimulation. John reports in his article that, after identifying the pattern for a specific stimulus, he was able to make an electrical replica of the pattern to stimulate cells where the electrodes were planted. It was found that by reproducing this pattern in the cat's brain or even only in parts of the cat's brain, the cat could be made to perform the response associated with the reproduced pattern even though the stimulus for a different learned response was being presented to it externally.

All of these findings led John to conclude that "cells combine to perform mental functions by a statistical process, and the average firing pattern controls the function."

> Each brain region has a characteristic signal-to-noise ratio for a particular operation. "Noise" refers to random cell firing, and "signal" to cell firing in rhythm with other cells performing the same operation. The more signal and the less noise, the greater the contribution of a given region of cells to a specific function. The regions conventionally thought to control a given function are actually those with the most signal and least noise for that function . . .
>
> One implication of this is that no one function can be attributed to the activity in any specific group of cells. Every mental operation, including consciousness itself is due to activity throughout the brain. (p. 52)

John's final statement is that "the memory of what is learned is not to be found in any specific brain region, but rather in its unique cell-firing rhythm . . . The brain's rhythms count for as much or more than the way it is put together."

2.3 The Cybernetic Brain

Although E. Roy John presents much new phenomena regarding neural function, and his statistical brain theory provides a much more elegant and intuitively satisfying explanation of memory than the connection theories, it still seems incomplete in many ways. The cybernetic brain model differs from the statistical brain model in that it does not separate the overall rhythms of neural firing from the structures by which they are propagated. It is the way in which the brain is put together that will for the most part determine the kinds of rhythms that are produced. Further, John seems to be limiting himself to only one level of learning and memory, a primarily cognitive level. To account for motor activity and coordination which result from cognition and association (that is, the facilitation or actualization of some learning or memory), one must take into account how learned rhythms will have an effect on brain structure: more specifically, how the larger changes in electrical potential, which John associates with learning, are rectified, filtered, and sorted in order to stimulate specific motor efferents in sequence to carry out a learned response. Another consideration would be how feedback may alter such sequences when some part of the environment is changed. What I am arguing against is the separation of mind and body.

In the cybernetic brain the concept of the localization of perceptual representations is not discarded, nor is the possibility of the equipotentiality of parts of the cortex for more than one kind of perceptual discrimination. These types of phenomena are considered in terms of the operation of the brain as a whole. The localization of certain types of perceptual or motor activity to certain parts of the brain is a useful way of coding information so that the visual aspects of an experience may be separated from auditory or kinesthetic input. Conceptualization, generalization, association, and the reaccess of sensory information, however, may take place on a qualitatively different level than the simple firing or refiring of specific neurons in specific parts of the cortex. In general, one might conclude that there are different hierarchies and levels of neural coding for different chunkings and communications of sensory information or processing. Different types of coding can occur in the interactions between (a) a neuron to itself (that is, individual properties of a neuron: size, shape, number of dendrites, firing threshold, degree of mylination, and so on) (b) a neuron to other neurons (c) a neuron and its environment (location in cortex or body, chemical composition of immediate environment, and so on) (d) different groups of neurons, (e) groups of neurons and their environment, and so on.

Appendix A

Electrostatic versus Electromagnetic Field

There are two ways of producing a measurable change in potential, electrostatic, and electromagnetic.

a. Electrostatic

Electrostatic phenomena are the result of the field associated with the charge of an electron. Electrostatic effects can generally be thought of as the result of the differences in concentration of charge between points in time and space. This type of effect is responsible for such physiological phenomena as:

1) the combinations of electrons with particles of opposite charge (protons) to produce atoms, molecules, substances, and so on;

2) interactions between ions and molecules due to respective valences;

3) concentration gradients between ions and molecules inside and outside of cell membranes;

4) potential gradients both within and exterior to neural membranes.

In fact, almost all of the recorded physiological properties and phenomena of brain function (chemical, biological, electrical) are explained in terms of electrostatics.

b. Electromagnetic

Electromagnetic effects are the result of a field propagated with respect to the rate of change of electrostatic potential (essentially, the movement of

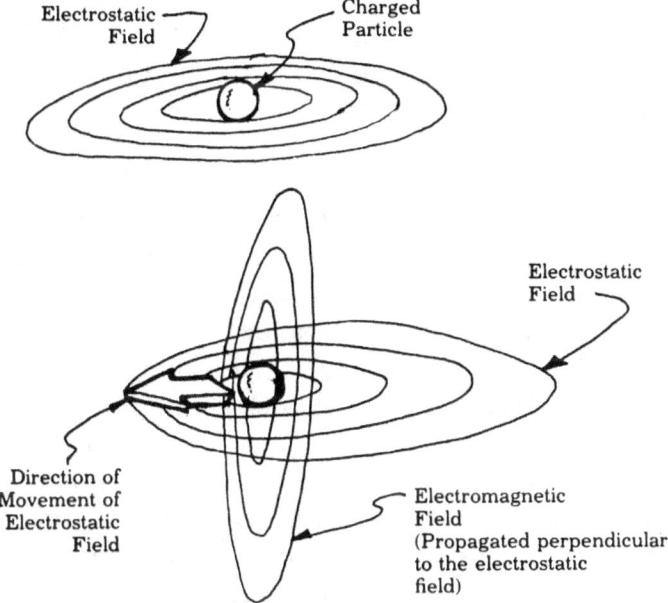

Figure 12.

electrons). An electromagnetic field is a second order potential generated in a direction perpendicular to the direction of an electron's movement. This field is thought to be the result of relativistic effects associated with velocity (particularly length contraction, so that there is a relative concentration of potential in a group of moving electrons other than that recorded as a result of their electrostatic fields). For example, the electromagnetic waves composing the spectrum of visible light are generated by the movement of electrons jumping from one orbit in atomic structure to another.

Because electromagnetic effects are the result of rate of change, one might predict that, in the consideration of the function of neural cells, cells or axons with a higher firing frequency and mylinated (insulated) cells or axons would tend to produce larger electromagnetic effects.

1. Field Effect

When an experimenter inserts an electrode into the brain of an animal, he/she is measuring the results of electrostatic field: the fields of the electrons in the area of the electrode affect electrons in the metal of the electrode, and so on. Changes in the chain of electrostatic events which is being observed, however, may be brought about by a number of events: (a) the relative concentration of positive or negative ions in the immediate area around the electrode, (b) changes in potential caused as the result of concentrations of electrostatic charge in surrounding areas (it is the summations and interference patterns occurring in larger areas as the result of these changes in concentration that E. Roy John hypothesizes about), or (c) changes in electrostatic potential brought about by electromagnetic field.

Most theories of mental activity seem to revolve around the lower order field effects used to describe particular chemical and biological interactions (the combination and interaction of chemicals and ions, membrane permeability due to transmitter substance, etc.), ignoring the possibilities of larger order field effects. The phenomena reported by E. Roy John seem to indicate neural responses resulting from these higher levels of activity: particularly (a) the recording of cells in different areas of the brain firing in the same rhythm to the same response, (b) random firing of cortical cells, (c) the elicitation of a predicted response from an animal by the external reproduction of a pattern of evoked potentials

(d) the ability of an animal's brain to integrate split externally-evoked potentials.

Many of these phenomena begin to make sense when explained in terms of higher level activity. Random firing can produce non-random effects in the larger overall system of the brain. Further, random firing may be the non-random result of other neural events.

Consider some of the possible results of field effect:

a) dielectric effects. Molecules form electric dipoles according to valence and the properties of the atoms of which they are formed. When placed in an electric field these dipoles will tend to line up in the direction of the field, to offset the effects of that field.

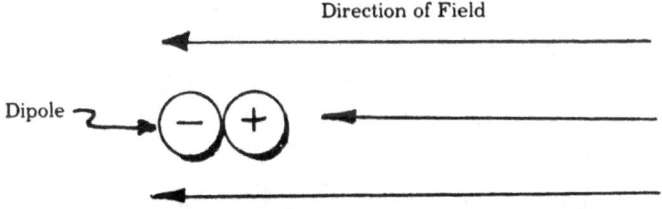

Figure 13. Dipole lining up with direction of electric field.

b) membranes have been shown to change their permeability constants when placed in an electric field. This is probably due to dielectric effects.

c) an alternating current is created by a direct current pulse (like those caused by the firing of a neuron) due to the expanding and collapse of the resultant electromagnetic field.

Keeping in mind these effects, one might postulate how a neuron in one part of the brain might fire in response to the firing of another, without there being any direct synaptic connection between the two:

(a) The membranes of different neurons have different firing thresholds and permeability properties. This is most obvious in receptors. Different receptors fire in response to different stimuli (some fire in response to pressure, others to temperature, others to the presence of certain chemical substances, and so on). The receptors in our eyes respond to extremely minute changes in electromagnetic fields in the form of light. One might postulate that in the cortex there is also a possibility of the occurrence of different membrane properties, perhaps corresponding to different layers of the cortex. The degree to which a neuron is mylinated might be considered one such property. The structure of the membranes of some cortical cells may be more sensitive to field effects than others.

(b) Depending on the structure, proximity or location (whether it is parallel or perpendicular, and so on) of one neuron with respect to another, it may come under more or less influence of the fields created, both electrostatically and electromagnetically, by the firing of that other neuron or group of neurons. These effects may make changes in the firing properties of the one neuron, perhaps causing it to fire in the absence of direct synaptic stimulation.

For instance, the rapid firing of a mylinated neuron or group of mylinated neurons may cause a response in a non-mylinated neuron in another part of the brain by generating an electromagnetic field which might 1) cause changes in membrane permeability (through the dielectric properties of the molecules which make up that membrane) beyond a certain critical point, and 2) evoke a change in the electrostatic potential (due to the expansion and collapse of the electromagnetic field) within that neuron, which may be enough by itself, or add up with other activity within that neuron, to release transmitter substance to another neuron.

Bibliography For Part I

1. Acosta, Cowan, Graham: *Essentials of Modern Physics.* Harper & Row Publishers, 1973.

2. Barret, Terrence W.: The Cortex as Interferometer: Transmission of Amplitude, Frequency and Phase in Cortical Structure. *Neuropsychologia* 7:135–48, 1969.

3. Bateson, Gregory: *Steps to an Ecology of Mind.* Ballantine Books, 1972.

4. Bateson, G.: "Reality" and Redundancy. *CoEvolution Quarterly,* pp. 132–135 Summer 1975.

5. Bingham, W. E. Jr.: Electromagnetic and Electrostatic Fields: A Neglected Area in Physiological Psychology. *The Journal of Psychology* pp. 225–31, 1954.

6. Callaway, E.: Schizophrenia and Interference. *Arch Gen Psychiat* 22: 193–208, 1970.

7. Cohen, D.: Magnetoencephhalography: Evidence of Magnetic Fields Produced by Alpha-Rhythm Currents. *Science* 161: 784–6 1968.

8. Geschwind, N.: Language and the Brain. *Sci Am* 226: 76–83, 1972.

9. Grinder, J., Bandler, R.: *The Structure of Magic Vol I and, II,* Science and Behavior Books Inc, 1975, 1976.

10. Grinder, J., Bandler, R., DeLozier, J.: *Patterns of the Hypnotic Techniques of Milton H. Erickson, M.D. Vol. I, II.* Meta Publications, 1975, 1976.

11. Gengerelli, J. A., Norman, J., Holter, Glasscock, W.: Magnetic Fields Accompanying Transmission of Nerve Impulses in the Frog's Sciatic. *The Journal of Psychology* 52:39–50, 1961.

12. Haley, J.: *Advanced Techniques of Hypnosis and Therapy.* W. W. Norton & Co., Inc. 1967.

13. Halliday, D., Resnick, R.: *Physics I and II.* John Wiley & Sons Inc, 1966.

14. Hubel, D.: The Visual Cortex of the Brain. *Sci Am* pp. 168–178 Nov. 1963.

15. John, E. R.: *A Model of Consciousness.* Unpublished, (1975).

16. John, E. R.: Switchboard versus Statistical Theories of Learning and Memory. *Science* 177:850–64, 1972.

17. John, E. R.: How the Brain Works—A New Theory. *Psychology Today* pp. 48–52, May 1976.

18. Just, A.: Hemispheric Specialization and Logical Typing in Aphasia: A Theory of Right Brain Communication. Unpublished (1976).

19. Kolin, A.: Magnetic Fields in Biology. *Physics Today* pp. 39–50 Nov. 1968.

20. Lamendella, J.: Some General Principles of Neurofunctional Organization and Some Manifestations of these General Principles in Second Language Acquisition. Preliminary Draft (1977).

21. Lamendella, John.: *Early Growth of Cognition and Language: A Neurophysiological Approach,* In Preparation (1977).

22. Leukel, Francis.: *Introduction to Physiological Psychology,* C. V. Mosby Co., 1972.

23. Miller, Pribram, et al.: *Plans and the Structure of Behavior,* Henry Holt & Co., Inc., 1960.

24. Milner, P. M.: *Physiological Psychology,* Holt, Rinehart & Winston, 1970.

25. Post, Steve: "Growth of Organic Systems," Unpublished (1975).

26. Post, Steve: "Thinking about Thinking: A Philosopher Gets His Hands Greasy," Unpublished (1977).

27. Pribram, Karl: *Languages of the Brain,* Prentice-Hall, Inc., 1971.

28. Pribram, Karl: *Holographic Hypothesis of Memory Structure in Brain Function And Perception,* Unpublished (1973).

29. Sperry, R. W.: "The Great Cerebral Commisure," *Scientific American,* Vol. 210, No. 1, Jan. 1964, pp. 42–52.

30. Suthers, R., & Gallant: *Biology: The Behavioral View,* Xerox College Publisher, 1973.

31. Thompson, Richard F.: *Introduction to Physiological Psychology,* Harper & Row 1975.

32. Watzlawick, Paul: *Pragmatics of Human Communication: A Study of Interactional Patterns, Pathologies, and Paradoxes,* W. W. Norton & Co., Inc., 1967.

PART II:
EEG AND REPRESENTATIONAL SYSTEMS

Robert B. Dilts, 1977

TABLE OF CONTENTS FOR PART II
EEG & REPRESENTATIONAL SYSTEMS

The Framework

I. The Neurology of Representational Systems	5
1. The Neuron	6
2. Anatomical Hierarchy of Neural Systems	8
3. Structure of Neural Logic	15
4. Structure of the Cortex	17
5. Transmission of Information in Neural Systems	20
II. Representational Systems and Behavior	21
III. Strategies	22
IV. T.O.T.E.S.	23
1. Nested TOTES	25
V. Generalizing Strategies	35
VI. The Neurology of Generalization	36
VII. Neurology of States of Consciousness	39
VIII. The Holographic Brain	42
IX. The Neurology of Learning	48
X. Neurophysiology of Consciousness	51
XI. Accessing Cues—Tuning in Brain Functions	52

The Experiment
  Purpose
  Equipment

Experimental Setup
Procedure
Results
Conclusions
Appendix A
Bibliography

THE FRAMEWORK

One of the most fundamental concepts of Grinder and Bandler's Neuro-Linguistic Programming model is that of *representational systems.* This concept begins with the fact that we, as human beings, do not operate directly on the environment in which we exist, but rather through sensory transforms of that environment that can be grouped into six major classes: vision (sight), audition (hearing), kinesthesis (tactile body sensations), proprioception (internal visceral and emotional states), gustation (taste), and olfaction (smell). All distinctions that human beings are able to make about their environment (internal and external) or their behavior must be represented in terms of these senses.

I. The Neurology of Representational Systems

In Neuro-Linguistic Programming, each sensory system is considered to be more than simply input mechanisms. They are also considered to be processing systems that initiate and modulate behavioral outputs. Each perceptual class forms a three-part (input-representation/processing—output) sensory-motor complex that is responsible for different classes of behavior. The first stage, *input,* involves gathering information and getting feedback from the internal and external environment. Stage two, *representation/processing,* involves the mapping of the environment and the establishment of decision making schema. *Outputs* are the causal transforms of the representational mapping process. It is this entire three-part network to which the term, representational system, applies.

Physiologically, a particular representational system is made up of a complex network of neurons, beginning with specialized receptors and sense organs located throughout the central nervous system. Peripheral receptors and transmitters will "fire," depending on their individual structural properties, to different types and thresholds of mechanical, electro-chemical, or electro-magnetic irritations in the individual's external and internal environments. These sensory distinctions are then transferred along neural pathways, in a fairly autonomous fashion, through the central nervous system, lower brain structures, and sensory relay nuclei. There are, of course, varying amounts of processing and outputs occurring at the various levels of consolidation along these pathways.

1. The Neuron

The basic carrier of information in the nervous system is, of course, the neuron: the transmitter and representational element of sensoral difference. The functional significance of a given neuron will be determined to a large extent by the genotype, its position with respect to other cells, its external and internal environment, degree of maturity, DNA content, types of proteins synthesized, membrane permeability, firing threshold, refractory period, volley firing properties, chemical transmitters (excitatory or inhibitory), sensitivity to ephaptic (electrical) as

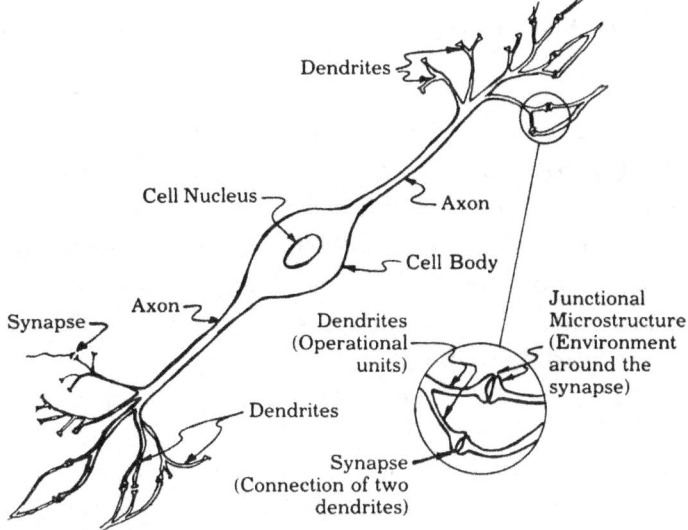

Figure 1. The Neuron: Basic carrier of information in the brain.

well as synaptic (chemical) impulse generation. The degree of *myelinization* of the neuron's axon, a fatty substance covering the cell which tends to speed up the transmission of impulses but reduce plasticity, will also affect functional significance.

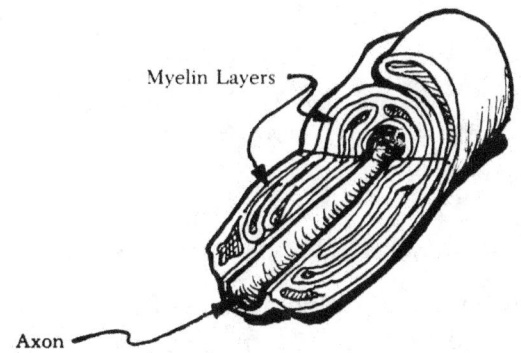

Figure 2. Myelinized neuron (section).

Neural networks are composed of two basic types of neurons, that can be found at all levels of neural systems, class I and class II. Class I neurons are large neurons with long myelinated axons that preserve the topographic arrangement of sensory receptors and motor effectors that they are connected to. The distribution of *class I* neurons seems to be specified during embryonic development, is invariant, and unmodifiable. They do not require sensory stimulation for their development.

Class II neurons are small unmyelinated neurons with short axons that integrate the activity of other neural elements, rather than function as receptors of effectors. *Class II neurons* require sensory stimulation for their development. The absence of such stimulation during the various critical periods of their development will produce physiological degeneration of the cell's activity along a number of parameters, including cell size and electric response.

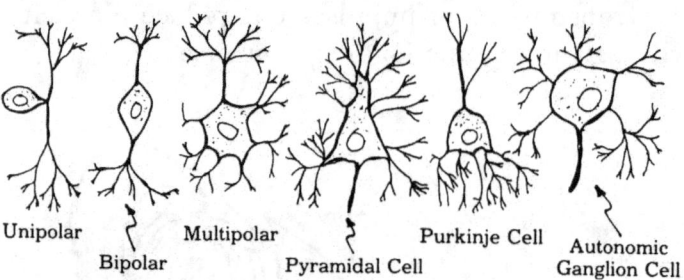

Unipolar   Multipolar   Purkinje Cell   Autonomic
   Bipolar   Pyramidal Cell   Ganglion Cell

Figure 3. Types of neurons.

2. Anatomical Hierarchy of Neural Systems

Spinal systems are composed of a preponderance of class I neurons and carry out little processing beyond the most basic reflex reactions.

Brain-stem systems are composed of both class I and II neurons and are basically concerned with gross motor coordination, sensory processing, and low level sensory-motor processing.

The *reticular activating system* is an internal core of neural tissue, extending from the spinal cord to the midbrain, composed of intermingled nuclei and fibers. The reticular core is surrounded by, and connected to, long afferent (input) and efferent (output) tracts of the brain stem and spinal cord.

As we move up the various levels of the nervous system, then, some of the sensory specific inputs are processed to certain degrees by class II neurons and the rest is transferred to higher level structures by class I neurons. The structural details and types of interconnections at each of these levels will determine the type of processing that takes place. The concentric arrangement of neurons within the nucleic structures causes different interactions and distributions of impulses than the layered levels of neurons in the cortex. In addition, each level also has a hierarchic relationship with structures at other levels. Lamendella (1977) outlines some of the properties of these hierarchies:

> Anatomically, the vertebrate central nervous system is organized into an ordered series of *levels*. Functional domains of information processing activity are realized in the nervous system as a hierarchical series of neurofunctional systems at increasingly higher levels, the hierarchy as a whole constituting a neurofunctional *metasystem*.
>
> (1) In general, functional systems and functional hierarchies have a bilateral representation in the nervous system: however, there also exist specialized forebrain systems that come to be *lateralized* only in the left or right cerebrum.

(2) A given system, subsystem, or metasystem may form part of more than one functional hierarchy and/or carry out more than one function.

(3) A higher level system within a hierarchy may stand in several different relationships with lower systems: (i) it may be simply *added in* to the information flow and work in concert with a lower system; (ii) it may *integrate* the activity of separate lower systems, thereby giving rise to new functional capabilities; (iii) it may *differentiate* or *specialize* the unified activity of a single lower system to produce novel functional capabilities at the higher level; and (iv) a higher system may be superimposed on a lower system.

(4) For a given functional hierarchy at any given moment, a particular level in the hierarchy may be said to be in *control* of particular types of input/output processing and the production of overt behavior within the functional domain of the hierarchy.

(a) *Prime control* over behavior may undergo rapid shifts up and down the hierarchy even while different aspects of functional control are distributed at several levels simultaneously.

(b) A higher system superimposed on a lower system tends to *inhibit* some subset of the lower systems functional activity, with the functions being retranslated at the higher level. In general, superseded lower systems tend to continue operating at this modified/reduced level of activity as a subordinate component of the functional metasystem.

(c) In certain contexts, a superseded lower system may retain the capacity to establish preemptive control over input/output functions in a domain normally reserved to higher system in the hierarchy.

(d) If a higher system within a hierarchy is incapacitated by disease or trauma, a superseded lower system may be released from its inhibitory influences and resume functioning approximately as it did before the higher system became operational. . . .

. . . An information/skill schema derived by a system at some level based on a non-wired-in interaction with the environment may be viewed as constituting an *infrasystem*. In-

EEG and Representational Systems

frasystems develop "horizontally" within a level, rather than "vertically" across levels in a functional hierarchy.

In terms of this vocabulary, what I identified earlier as a representational system would constitute an infrasystem within the neural hierarchy that keeps an informational integrity across various levels of processing. Sensory-specific, stimulus-bound information would be transferred across levels by class I neurons. Crossover functions, such as integration and differentiation, would take place via class II neurons.

Each representational system would input a particular class of information into each level of the system, and the crossover connections within and between each infrasystem would determine the kind of information output. Differences in information input will cause different output effects. Outputs are

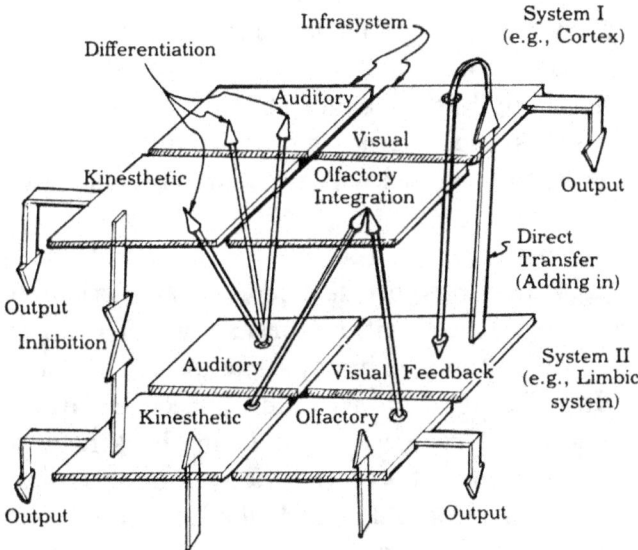

Figure 4. Neurological Metasystem

directed toward the external environment through motor efferent tracts, and to the internal environment through the hypothalamus and the autonomic and endocrine systems.

The *autonomic nervous system* regulates the activity of smooth muscle tissue of the visceral organ system and controls such functions as digestion, elimination, cardiovascular activity, reproduction, respiration, glandular activity, and some peripheral involuntary muscles.

The *endocrine system* controls vegetative and metabolic functions, and glands producing hormones.

Hypothalamic nuclei regulate intake of food and water, blood pressure, body temperature, water excretion, salt balance, and pain and pleasure functions.

These systems are structures that tend to be specialized for outputs that also occur at various levels. The regulatory functions mentioned above are mediated by structures in the *limbic system,* which is at a structurally higher level than the reticular formation. At this level of processing, most of the functions still seem to be "hard wired" (primarily class I neurons).

Motor responses also occur at various levels of processing, becoming more detailed and sophisticated at each level, but less critical for survival. The *cerebellum* is probably the primary structure for motor output. It regulates balance and the coordination of movements. It receives input from the kinesthetic muscle receptors, the vestibular system of the inner ear, and forms a negative feedback loop from the premotor areas of the cortex to the cerebellum and back to the motor cortex, that allows for finer coordination of movement. Outputs from the *cerebellum* regulate muscle tone and postural adjustments in the skeletal tract.

The *thalamic nuclei* are also contained within the limbic system. These nuclei relay projections of receptor topography for all sensory modalities except olfaction (which is projected directly to a lower cortical area). As well as simply consolidate and transfer information, some thalamic structures tend to send a qualitatively different output to the next level. The *specific thalamic projection system* emits topographically organized sensory output for vision, audition, and kinesthesis to topographically specific areas of the cortex. The *diffuse thalamic projection system* has a more widespread distribution and less modality-specific projections to secondary sensory areas of the cortex. Areas supplied by the diffuse thalamic projection system become more regular and synchronized and tend to have a larger amplitude of firing. These *slow potentials* are thought to be a recruiting response caused by excitation reverberating (through feedback interactions) back and forth between thalamus and cortex. These slow potentials will be discussed in more detail later.

Sensory information, then, is projected into various cortical areas with varying degrees of modality-specific activity. The density distribution is such that certain areas of the cortex tend to be specified for the majority of one type of sensory activity. For instance, the majority of visual activity takes place over the occipital lobes of the human cortices, the majority of auditory activity tends to occur within the temporal lobes, the majority of kinesthetic activity (tactile body sensations) within the parietal area, and the majority of motor activity tends to be consolidated in the precentral cortex. These areas have been iden-

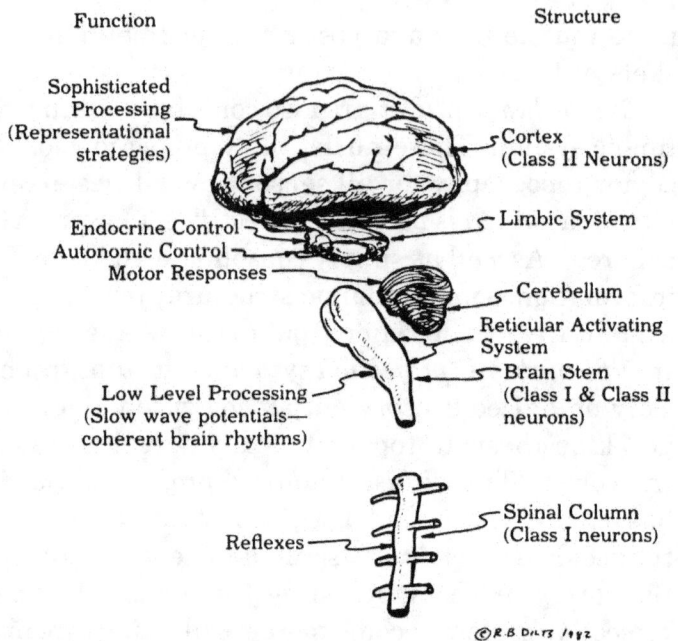

Figure 5. Anatomical hierarchy of neural systems (schematic).

tified by measurement of neural electric activity in response to sensory-specific stimuli and tests involving insult, destruction, and direct electric stimulation of the various areas in question.

There is, as we pointed out earlier, considerable overlap in the cortical representation of sensory information in some areas of the cortex and even in the limbic system (E. R. John, 1973, 1976; Bach-y-Rita, 1974). Cells in many areas of the cortex and the limbic system fire at the same time and in the same rhythm to a particular sensory stimulus. The importance of this redundancy of sensory information will be discussed later in reference to synesthesia patterns, parallel processing, and the development of representational system primacy.

3. Structure of Neural Logic

The cortex is the highest structural level within the neural metasystem, and is where the most sophisticated processing occurs. The various layers and areas of the cortex differ in the types of cells found in the layers, the thickness of the layers, the number and types of sublayers, cell density, number and types of interconnections. The interconnections of class II neurons may be of a number of types, in many ways similar to that of a digital computer. First, the synaptic junction may be of an excitatory or inhibitory nature, chemically. An *excitatory* junction would increase the probability of two cells transferring a particular impulse. An *inhibitory* junction would decrease this probability. There are of course varying degrees of excitation or inhibition within any particular number of synapses. Karl Pribram (1973) calls this degree of interaction the *coupling coefficient* for a particular synapse.

Second, each neuron will have a multiplicity of synaptic junctions, each of varying coupling coefficients of an excitatory or inhibitory nature. The temporal and spatial summation of activity at these synaptic junctions will also contribute to the firing properties of a particular neuron. For example, consider the following:

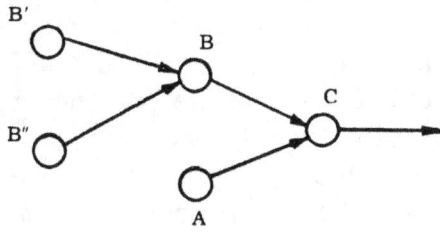

Figure 6.

Each circle in the diagram represents a particular neuron: A, B, B', B", and C. The arrows indicate the direction of an impulse along a dendrite with the tip of the arrow being the synapse. B' and B" are inputs to B, and B and A are inputs to C.

If A and B are both excitatory inputs to C and are of high coupling coefficient, then if either A *or* B were to fire, C would fire. This is called an *Or Circuit* in computer terminology.

If A and B are both excitatory inputs to C and are of low coupling coefficient, then A *and* B would both need to fire together to initiate activity in C. This is known as an *And Circuit*.

If A were an inhibitory input and B an excitatory input, then in order for C to be initiated A would have to *not* fire *and* B would have to fire. This is called a *Nand Circuit* or gate: meaning not-and. This is, of course, provided that B is of sufficiently high coupling coefficient to initiate C on its own, and that A is of sufficiently high coupling coefficient to inhibit B from initiating A.

If A and B were both inhibitory synaptic junctions then neither A nor B could be firing if C were to be initiated. This *Nor Gate,* of course, requires that C would have to normally be in a firing state or that it would be receiving some other excitatory stimulus from some other neuron or its environment.

Each of these logic functions would also operate on B as a result of inputs from B' and B". So that B' and B" could form an excitatory And gate for B that, when initiated, would act to inhibit C.

It is undoubtedly through this type of circuitry that many of the complex neural input-output, stimulus-response chains are organized. The hierarchic levels of structures we described earlier are composed of complex networks of these digital computational circuits that form input-output transform planes and feedback loops. Simple behavioral reflex arches would be composed of such chains, involving primarily class I neurons. Learned behavior would be the result of developed integration of activity by class II neurons.

In addition to the development of these four types of neural gates, the functional significance of a given neuron will be determined by its genotype (size, shape, number of dendrites, and synapses; and so forth), its position with respect to other cells (the number of cells in its immediate environment), degree of maturity, DNA content, types of proteins synthesized, membrane permeability, firing threshold, refractory period, volley firing properties, sensitivity to ephaptic (electric) as well as synaptic (chemical) impulse generation, and the degree of myelinization.

4. Structure of the Cortex

The layers of the cortex vary in all of the aspects mentioned thus far. The white matter of the cortex is composed of myelinized class I, stimulus-bound neurons (the myelin is a white fatty sheath). These all run parallel to the surface of the cortex. The gray matter of the cortex is composed of predominantly class II neurons, which tend to form six layers, each differing in the size of neural cells and number of dendrites, which also run parallel to the surface of the cortex. The lower layers of the cortex tend to have primarily large neural bodies with fewer fine dendritic connections. These bodies become progressively smaller as they develop into the upper layers, until they reach the top layer, which is primarily composed of dendritic fine fibers. These fine fibers tend to conduct impulses more slowly than the larger axons of lower level neurons. Their firing patterns tend to be more coherent and stable.

Stimulus-bound information is provided to the cortex by varying densities of class I neurons which run perpendicular to the surface of the cortex from the white matter.

Depending on its functional significance, the constitution of the cortex will vary over different areas. Primary sensory cortex, for instance, is primarily

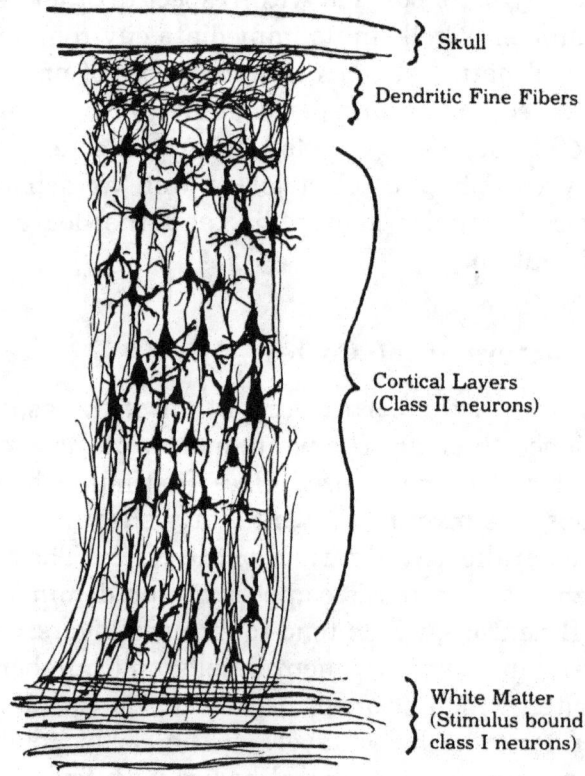

Figure 7.

composed of cells that are granular, thinner, and packed close together. Cells in the motor cortex are agranular, thick, and large.

It is interesting to note that there are three motor-body projections in the cortex. The *primary motor area* is in area-4 over the *central* lobe of the cortex. These cells control movements on the side *contralateral* to the cerebral hemisphere that they are in. A second projection is in the *sylvian fissure*. This tends to be a mirror image of the area-four projection, but actuates activity on the *ipsilateral* side of the cerebral hemisphere in which it is located. The third area is what is called the *supplementary motor area*. It is located in the area from the *longitudinal fissure* to the *cingulate gyrus*, and has a *bilateral* effect on movement.

The existence of these three motor areas, or infrasystems, is important in that it proposes the possibility of the development of functional prime control of separate motor schema by different representational infrasystems.

The establishment of prime control by a system or infrasystem over any functional domain is the result of signal-to-noise values of activity within the specific functional areas of the neural metasystem. Signal would be determined by the relative amount of system-specific activity within the area. For example, the occipital area of the cortex would have the highest signal value for visual information of the cortical areas.

Similarly, within a hierarchy, a higher or lower level system may take prime control depending on the amount of activity within that system.

5. Transmission of Information in Neural Systems

Signal values are established in two major ways: structurally and dynamically. I have already talked at some length about structural aspects of neurofunctional systems. The dissemination of information structurally takes place mainly through the convergence and divergence of nerve impulses onto and from a neuronal pool. This is a purely structural interaction and takes place via the effects of synaptic interconnections between neurons. Excitatory and inhibitory transmitters are released at the synaptic junctions between cells. This causes a change in the permeability of the cell's membrane to certain chemicals in its immediate environment, resulting in a rapid influx and outflux of sodium and potassium ions. This activity causes changes in electric-field potential near the cell (due to the movement of chemical ions in the environment) and, depending on the firing rate, causes the firing neuron to also release chemical transmitters to other cells at its own synaptic junctions. We have called the type and sensitivity of the biochemical connections between the synaptic junctions of two neurons the *coupling coefficient*. Coupling coefficients are established between both class I neurons, (which transmit stimulus-bound information), that tend to be organized in verticle parallel columns, and class II neurons (which integrate and process stimulus-bound information) that conduct impulses horizontally, perpendicular to class I neurons. Class II neurons have many thin dendritic fibers that interconnect with other cells. These fine fibers tend to conduct impulses much more slowly than the thick myelinated axons of class I neurons and probably contribute a great deal to coherence between the rapidly fluctuating activity within neural networks.

Signal within these structural elements will be determined by such properties as density of neurons and dendrtites, amplitude and frequency of neural firing, coupling coefficients, and the internal circuitry of a particular input-processing-output networks (that is, And, Or, Nand, and Nor gates). These values will to a large extent be determined by the strength of the input and the types of feedback wired into the system.

II. Representational Systems and Behavior

Depending on its signal value, then, a particular representational infrasystem may, at various levels in the hierarchy, establish prime control over a particular functional domain. Bandler and Grinder (1975, 1976) have identified, in their clinical work, an individual's most *highly valued* or *primary representational system,* as they call it. This was based on their observation that many individuals tend to rely heavily on one particular domain of sensory experience to serve as a primary guide for their behavior. The establishment of a most highly valued representational system leads to shared personality and behavioral characteristics (for such properties as talents, interests, values, beliefs, motivation, learning, and other organizational patterns) for individuals of similar representational types.

Engineers, painters, and chemists, for instance, may be more visually oriented; athletes and dancers more kinesthetically oriented; musicians and poets more auditorily oriented. Different variations within a particular field also tend to be determined by different representational orientations. In the field of psychotherapy, for example, different therapies are geared to different representational systems. Art therapy and Jungian symbol analysis tend to be more visually oriented; massage, bio-energetics, and other body therapies would be more kinesthetically oriented; gestalt and Riechian therapies rely on emotional expression; Freudian and Rogerian techniques tend to be auditorily oriented.

Representational prime control is undoubtedly a function of task. Learning a musical instrument requires the establishment of classes of perceptual-motor programs that are different from those utilized for gymnastics or mathematics. Each of these programs will contain steps in which a representational system, other than the one identified as most highly valued for the task, may have prime control at any point in time. Perhaps a more relevant concept to introduce at this time, then, would be what, in the Neuro-Linguistic Programming model, is called a *strategy*.

III. Strategies

A strategy is a sequence of representational activity that leads to a behavioral outcome. Thoughts, for instance, are strategies composed of varying sequences of internal visual images, sounds, voices, internal dialogues, and feelings. The two most important aspects of a strategy are: 1) the representational system in which the information at each particular step is coded, and 2) the sequential order and relationship between representations.

For example, a particular strategy for creative dance may go something like: step 1) the individual hears music from an external source; step 2) the individual gets an emotional feeling from the music; step 3) the individual constructs a visual image of him/herself moving with various tempos and movements in relation to the feeling; step 4) the individual transforms the movements from the image into actual kinesthetic body movements.

A representational system's analysis of this strategy would show that it initiates from an external auditory input. This, in turn, initiates internal visceral feelings, the feelings initiate internal visual imagery, which, then, initiates motor kinesthetic output. In general, for a strategy analysis, the last step before the motor output is considered most highly valued for that task. Most effective strategies occur in the form of a TOTE.

IV. TOTES

The TOTE is an extension of the "reflex arc" (a stimulus-response chain) in behavioral theory. The TOTE model was originally proposed by George Miller, Eugene Galanter and Karl Pribram in their book *Plans and the Structure of Behavior* (1960). The term, which stands for the series Test-Operate-Test-Exit, is a neurological model for the formal sequence of the internal processing of some stimulus. It is an extension in that the TOTE incorporates a feedback operation within the chain of behavior, itself, rather than considers it as a single linear action. As Miller, Pribram, and Galanter explain:

"The test represents the conditions which have to be met before the response will occur."

If the conditions of the *test* phase are met, the action initiated by the stimulus *exits* to the next step in the chain of behavior. If they are not met, there is a feedback phase in which the system *operates* to change some aspect of the stimulus or the organism's internal state; after which another *test* is made. Miller et al. write:

"Then the response of the effector (the output neuron) depends upon the outcome of the test and is most conveniently conceived as an effort to modify the outcome of the test. The action is initiated by an "incongruity" between the state of the organism and the state that is being tested for, and the action persists until the incongruity is removed. The general pattern of reflex action, therefore, is to test the input energies against some criteria established in the organism, to respond if the result of the test is to show an incongruity, and to continue to respond until the incongruity vanishes, at which time the reflex is terminated. Thus there is "feedback" from the result of the action to the testing phase, and we are confronted by a recursive loop."

The assertion of the concept of the TOTE is that the operations (behaviors) that an organism performs are constantly guided by the outcomes of various tests. Miller et al represented the TOTE visually as:

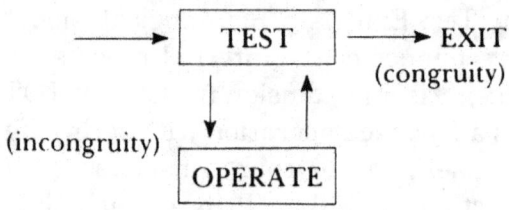

Figure 8.

The test phase may involve something as simple

EEG and Representational Systems

as a threshold test. That is, the impulse must be above or below a certain intensity value before the TOTE will exit to the next program. If it is not, the organism may do something (operate) to increase or decrease the intensity of the impulse. When you adjust the volume dial on your radio or stereo, you are performing a TOTE of this type. As you turn the knob, you continually test the sound volume by listening to it. If the volume is too low, you operate by turning the knob clockwise. If you overshoot and the volume becomes too loud, you operate by turning the knob counterclockwise to reduce the intensity of the sound. When you have adjusted the amplifier to the appropriate volume, you exit from the "volume-adjusting" TOTE and settle into your comfortable armchair to continue reading. We can illustrate this example of the TOTE process in the following way:

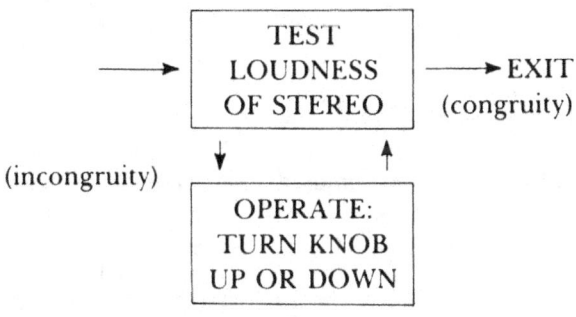

Figure 9.

1. Nested TOTES

One important characteristic of TOTEs is that the operate phase of one TOTE may include a number of other tests and operations: that is, the operate phase of one TOTE may include other TOTEs.

TOTEs may have a hierarchic arrangement, then, with respect to one another so that one TOTE may be nested within the operate phase of another. The example of this process given by Miller et al. is that of hammering a nail.

A carpenter, for example, may have a very abstract level TOTE that we will call "making a table." The test phase for this TOTE may involve a number of subroutines or subTOTEs including: "attaching legs to table surface." This TOTE may be composed of other subTOTEs including "hammering nails through a table surface to the legs." The test for such a TOTE may be that the nail head is flush with the table surface. If this test is not satisfied, the carpenter may need to go through an operate phase, hammering, which involves two sub-TOTEs: "lifting hammer" and "striking nail." The specific TOTE sequences of hammering a nail may go something like the following description given by Miller et al.:

> "If this description of hammering is correct, we should expect the sequence of events to run off in this order: Test nail. (Head sticks up). Test hammer. (Hammer is down). Lift hammer. (Hammer is up). Test hammer. (Hammer is up). Strike nail. Test hammer. (Hammer is down). Test nail. (Head sticks up). Test hammer. And so on, until the test of the nail reveals that its head is flush with the surface of the work, at which point control can be transferred elsewhere. Thus the compound of TOTE units unravels itself, simply enough, into a coordinated sequence of tests and actions, although the underlying structure, which organizes and coordinates the behavior, is itself hierarchical and not sequential."

The series of "nested" TOTEs described above is diagrammed in figure 10.

The Neuro-Linguistic Programming model, through the concept of representational systems and cognitive strategies, extends the TOTE model an-

EEG and Representational Systems

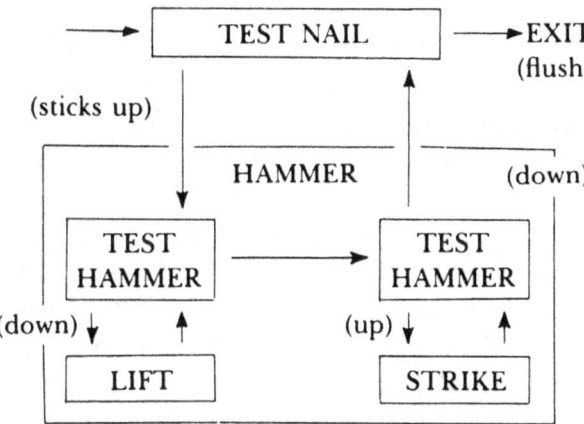

Figure 10.

other step. The Neuro-Linguistic Programming model asserts that, for higher cognitive processes and complex behaviors, the test conditions of the TOTE and most of the operations must take place through one or a combination of representational systems: that is, the incoming stimulus is tested against some stored representation.

For instance, in the above example of hammering a nail, the test of whether or not the nail is flush may be given by comparing the incoming visual experience of the position of the nail (we will abbreviate *visual* experience that comes from *external* sources as V^e) with some stored visual representation of what a nail looks like when it is flush (*internally* generated *visual* experience will be abbreviated V^i). We will show this comparison as V^e/V^i. The same comparison may also be made through tactile kinesthetic experience—the feelings in the carpenter's hand and arm will feel different when he hits a nail that is flush from when it isn't. The representations to be compared here are kinesthetic external versus kinesthetic internal, or K^e/K^i. The comparison may also be made auditorily, because the sound of the blow of the hammer against the nail will sound differently when the nail is flush. This comparison would be auditory external versus auditory internal, A^e/A^i.

A good carpenter can probably make the test very easily and comfortably with any of the representational systems; an inexperienced carpenter may not be able to. The concept of representational system primacy (see, *The Structure of Magic II* and *Patterns I and II*) asserts that many individuals tend to value one representational system over the others to perform their tests and operations. A visually oriented individual, for instance, would always look at the hammer and the nail to test them; a kinesthetically oriented individual would do it by feel.

The experience of *congruence* or *incongruence* as a result of a test may also be represented. When the carpenter tests the nail, for example, and notices that it is not flush, the incongruence between the external experience and the stored experience may be represented through an image, sound, voice, or feeling. He may hear a voice in his head saying, "No, it needs more hammering," he may get a feeling that initiates the hammering subroutine, he may hear a discordant sound, or see an internal image of the hammer pounding. The system that performs this function is sometimes known as a *reference system* in NLP. We show a kinesthetic representation of an incongruence between two visual representations as:

$$V^e/V^i \rightarrow K^i_{(-)}$$

A test need not take place only between external and internal representations. They may also take place between internal representations. They must, however, remain in the same representational class. That is, in general, a visual representation must be tested for congruence with another visual representation, auditory must be with auditory, kinesthetic with kinesthetic, and so on. I would postulate that the simultaneous pairing of representations within the same representational system is a function of the two cerebral hemispheres in human beings.

Operate procedures are generally the result of synesthesias (neural crossover connections) between representational systems, or the motor system in the case of overt behavior.

By making the representational form of test and operate procedures explicit, through strategies, the NLP model makes the transfer of any behavior that may be broken down into TOTE units more accessible and systematic. This is useful in the therapeutic context, because it makes patterns of behavior (whether they are problems or resources) more understandable and controllable. It is also useful in law, business, medicine, and education, and anywhere that face-to-face communication is involved.

Let us consider a fairly simple example, from education, of two different spelling strategies: the visual speller, and the phonetic speller. We have observed that a visual speller, when presented with a word (an external auditory stimulus), will go through some synesthetic operation procedure that transforms the word into a constructed visual image of the entire word. This constructed image is then tested against a remembered visual image of the spelling of the word. The congruence or incongruence of the two spellings is represented as internal kinesthetic feelings—if the constructed image does not "look" right, the speller gets a negative feeling and operates to construct another image. If the two images are congruent, the individual feels good and exits to a TOTE in which the image is vocalized. The operation phase will most likely be a synesthesia in which the word is repeated internally and another image is constructed. A second common operation is that of writing the word down externally. This alleviates the effort of holding the constructed image in the mind's eye and frees the brain to concentrate on finding the appropriate remembered reference image. The TOTE continues until an image is generated which, when tested, initiates a positive feeling.

This TOTE of a visual speller is represented in Figure 11A on the following page.

The phonetic speller has an operation phase in which the word is sounded out: that is, the individual slowly repeats the word (internally or externally) and makes a visual image of each phoneme. Once a complete visual image is made, the individual reads it to him/herself and tests the sound of the word she/he reads against the sound of the word originally presented. If they sound the same, the individual may say to him/herself, "that's it." If there is an incongruence, they may say, "no, try again," and cycle back through the operation. The operation phase described above undoubtedly consists of a number of sub-TOTEs. For the purposes of this analysis, however, we have chosen to represent the following as a single operation: auditory (break up into phonemes): visual (visualization of phonemes): auditory (pronounciation of complete image).

EEG and Representational Systems

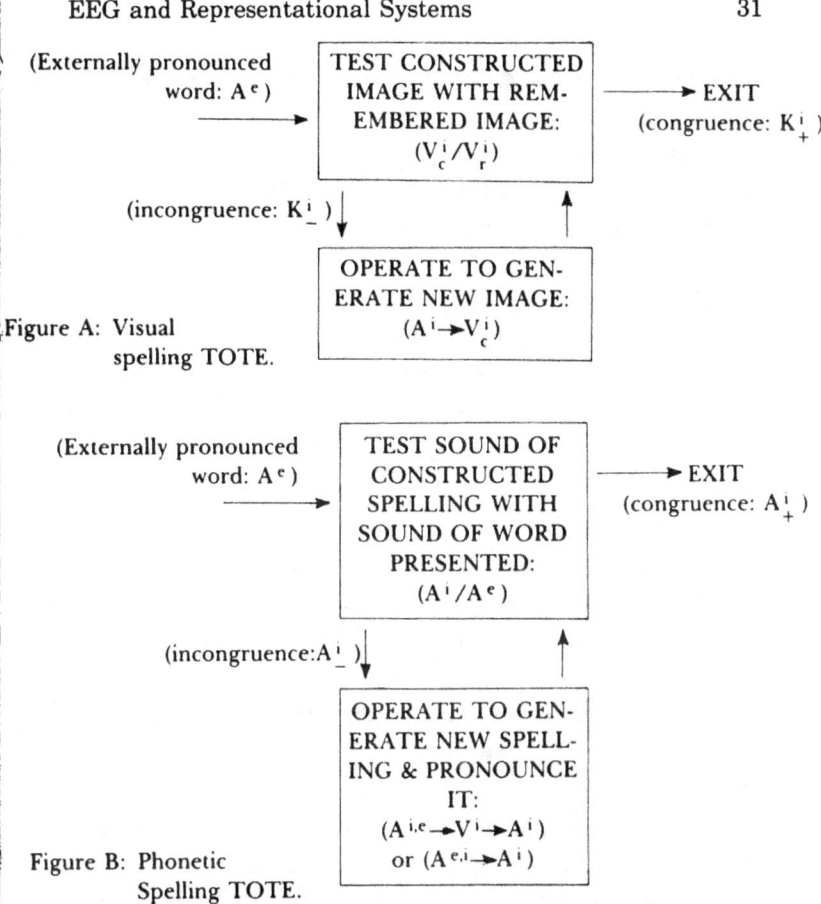

Figure 11.

It has been our experience that, since the visual coding of the English language does not follow phonetic rules, individuals with a visual strategy tend to be better spellers. "Thier," "their," "there," and "they're," for example, may all be pronounced in the same way, although, visually, they are very different. In most educational institutions children are not taught the formal aspects of learning when they are taught to spell; they are simply given feedback for whether they have spelled correctly or incorrectly, according to some standard. The children are left on their own to come up with a strategy (those described here are only two of an infinite number of possible permutations) that may or may not be the most effective or efficient for the task of spelling. Notice that, even if someone was able to become accurate at phonetic spelling, sounding out each word as you edit a paper or manuscript would be an incredibly laborious task.

By making these formal steps explicit, through the NLP model, we hope to increase the effectiveness of education in all disciplines. Different kinds of tasks involve the utilization of different representational systems in different sequences. By using the model of NLP, these sequences may be explicitly formalized.

The goal of representational analysis is twofold: 1) to find the most appropriate representational systems for the steps that lead to a particular outcome (such as using the visual representational system for testing in spelling); and 2) to be able to use all representational systems as resources for learning and performing. This means that in cases where more than one representational system may be used for the task (such as in the hammering example), the individual has the choice to substitute another representational strategy, should it be appropriate for the context. This serves to greatly expand one's repertoire of choices and abilities.

We have found TOTE diagrams to be somewhat laborious and impractical as a notational format for

EEG and Representational Systems

strategies. In NLP, we have streamlined the TOTE representation into a linear string of representations. Thus, we would notate the TOTE in Figure A as:

$$\underbrace{A^{e,i} \rightarrow V^c / V^i \rightarrow \underset{(-)}{K^i}}_{} \underset{(+)}{\rightarrow} \text{Exit}$$

This shows that the individual begins auditorily (the e,i superscript indicates that it may come from either external or internal sources), derives a constructed internal visual image, and tests it against a remembered image. If the images match, the strategy exits; if not, a negative feeling causes the individual to try again. This representation carries most of the information shown by a TOTE diagram. When it is important to show the hierarchic nature of recursive subroutines we notate them as follows:

$$A^{e,i} \rightarrow \begin{matrix} V^i_1 \\ V^i_2 \\ V^i_n \end{matrix} \rightarrow K^i$$

This sequence would indicate that the individual generates a series of visual images from the auditory experience and then weighs them kinesthetically, possibly by choosing which one feels the best.

The phonetic TOTE in Figure B would be shown as:

$$\underbrace{A^{e,i} \rightarrow V^c \rightarrow A^c/A^r \rightarrow \underset{(-)}{A^i}}_{} \underset{(+)}{\rightarrow} \text{Exit}$$

This shows that the individual makes an image out of the auditory stimulus and then tests it auditorily (by sounding it out). These two notational sequences show the basic differences in the two strategies.

There are, of course, a number of levels of detail in strategy analysis, and, as a general rule, the Neuro-Linguistic Programmer will choose that level of detail that most suits the situation and outcome he/she is dealing with.

It should be noted that a test need not always involve the comparison of two representations, but may be a test of the intensity of the representation. That is, the feeling, sound, or image may need to reach simply a certain threshold value, due to the assistance of the operation procedure, before the TOTE will exit. In this way, a stimulus in one representational system may be tested against a representation in another representational system, although the nature of the congruence or incongruence will be different. An image may need to be stronger than a certain feeling, for example, in order for the TOTE to exit.

Considered in terms of the physiological processes discussed in this paper, we may say that threshold tests will be the result of the interaction and effect of the channeling of stimuli into intensity nodes by the synaptic gates described earlier. Comparison tests may be the result of interference patterns (which would show congruence or incongruence) between impulses generated as the result of coherent input (such as alpha) intensities.

We could show a simple TOTE as Figure 12.

Here, neurons A_1, A_2 and A_3 are all providing stimulus input of an excitatory or inhibitory nature to node B. B has an excitatory connection to node E, and an inhibitory connection to node C. So that when B fires, E will be excited (the TOTE will exit) and C will be inhibited (the operation procedure will cease). The test occurs as the inputs from B' and B" interfere (positively or negatively) with those from A_1, A_2 and A_3. As long as B does not fire, the C-D operate circuit will continue to alter the input of A_3. (The C-D cir-

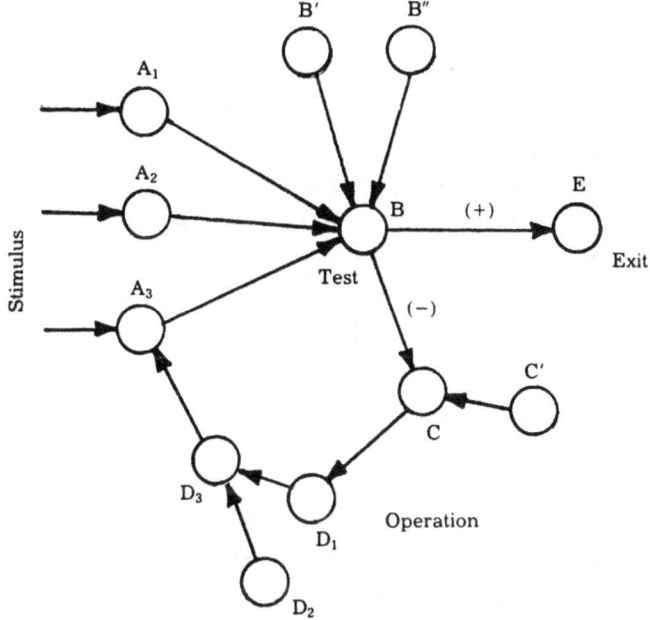

Figure 12. Neurological T.O.T.E. structure.

cuit could just as easily be altering the input of B' and B".) This continues until the right combination is found and B fires, inhibiting the C-D operation.

Input, feedback, operation, and exit phases of TOTEs may occur within infrasystems at the same level or between those of different levels in the hierarchy. An operation, for example, may be the function of a higher or lower system depending on the function of the operation or TOTE.

V. Generalizing Strategies

One important aspect of strategies is that they are strictly formal: that is, the same sequencing of representational systems may be applied to a number of different types of behaviors. In listening to a lecture, for example, an individual may apply the same strategy he/she uses for dancing. The individual may listen to the external words of the lecturer, the tonal and verbal aspects of the words initiating visceral feelings. From these feelings the individual constructs an image of him/herself participating in what the speaker is talking about, or constructs an image of something that speaker is describing. These images are then translated to movements, either minimally as the individual fidgets in his/her seat, or through internally generated body sensations.

Another individual might reverse two of the steps in the strategy. He/she may hear the words externally, make internal images in direct response to the words, which, in turn, initiate internal feelings (in response to the images). Motor kinesthetic output would then be initiated on the basis of these feelings. This simple change in strategy may make a great difference in the eventual behavioral outcome.

VI. The Neurology of Generalization

The phenomenon of applying a strategy to contexts outside of the one in which it was initially developed is called *Learning II* or *learning to learn*. This may be understood neurologically by considering the following: Lamendella (1977) identifies three major types of genetically programmed cellular migration within neural systems:

> "1) For some cells there is random migration and selective trapping of the cells, with their subsequent shaping by the environment which trapped them, (2) for other cells there is random release followed by selective migration to particular locations, and (3) the sequential origin and programmed release of cells destined for specific locations." (p. 36)

The trapping and shaping of neural cells is a function of the cell's environment. As was pointed out earlier, the presence or absence of sensory stimulation during various critical periods in the growth of class II neurons can cause development or degeneration, respectively, of the cell's activity along a number of parameters including cell size and electric response. The amount of time an individual spends utilizing a certain strategy at various critical periods, then, may cause the strategy to consolidate neurologically as cells migrate to form the various interconnected circuit gates. This process also, undoubtedly, contributes to the development of a most highly valued representational system.

Such processes as the cell migrations, described above, contribute to the establishment of structural signal values and prime control mechanisms. The second major mechanism for the distribution of signal values in neural systems is ephatic (electric) rather than synaptic (biochemical). Changes in electric field are caused by the firing of interacting groups of neurons, and the changes in the ionic density of the cell's environment, caused by neural firing, affect the chemical and electrical content of the cell's environment. Pribram (1974) calls the environment around two interacting neural processes their *junctional microstructure*. Changes in junctional sensitivity caused by neural activity outside of the immediate environment of a particular cell, or by the introduction of foreign chemicals through the blood system, can alter the biochemical structure of the cell membrane and the existing density of ions in the cell's environment. These changes will affect the neuron's resting potential, firing threshold, rate of firing and conduction, and so forth.

Because they are not insulated by myelin, class II neurons are much more sensitive to changes in the electro-chemical state of their environment. State dependency of class II integrators is important, because it helps make learning (the establishment of strategies) context sensitive, and also allows for plasticity and flexibility within the structural disbursment of signal values that establish prime control. A change in the state of a class II neural network can change the pre-established signal channeling properties of the circuitry.

Considered in terms of the hierarchic neural metasystem discussed earlier, the lower level regulatory, reflex, autonomic, and endocrine systems are more "hard wired" and innate, and less sensitive to minor changes in the neural environment than higher level systems. They react adaptively (see Ashby, 1959) to keep the individual's internal state in equilibrium. It is, of course, important to the survival of the organism to have such regulatory functions prewired and hard programmed. This is not to say that autonomic and endocrine functions are not affected by state or inputs and feedback from other systems and infrasystems. Many receptor inputs to these systems are, in fact, triggered by changes in chemical environment in various areas of the body. Further, higher level activity can act to enhance and inhibit lower level regulatory drives, and so forth.

In general, hard wired behavior at lower levels is more specified for particular inputs. It is less flexible with respect to behavioral triggers than higher order, non-regulatory behaviors which may have multiple derivations across and within interacting representational infrasystems, in the form of various strategies. Higher order cortical functions (strategies) serve to organize and sequence representations to allow for more complex and sophisticated coordination of output and tend to be more affected by ephatic states. Pribram (1976), for instance, reports that even extensive removals of motor cortex in monkeys did not impair movement within the monkeys' repertoire, but rather had a great effect on the coordination of *sequences* of movements. Anyone who has had experience with psychogenic drugs or behaviorally-induced altered states of consciousness will probably recognize that while his/her typical strategies for correlating experiences (making meaning) were interrupted, important regulatory functions remained intact.

Ephatic states will be determined by such parameters as rates of conduction of local neurons and dendrites, signal values of local neural networks (which effect the ionic density of the junctional microstructure), and, it is important to note, by inputs from lower level structures. The input from the diffuse thalamic projection system, for instance, will greatly effect the state of the cortex. Similarly, feedback from higher level structures will effect the state of lower systems.

VII. Neurology of States of Consciousness

An understanding of the properties of neural states will greatly enhance one's understanding of neurologic functioning and the interpretation of the EEG research to be discussed in the second half of this paper. EEG surface electrodes obviously pick up much more information about neural (particularly cortical) states than they do about the activity of any individual group or network of neurons. It is at this level of detail, then, that the data must be interpreted.

The first insight into the functional significance of cortical ephatic states was provided by Barrett (1969), who likened the cortex to an interferometer. He proposed the following: *cortical-slow potentials*, determined by the coherence of inputs from lower structures and the slower conduction rates of dendritic fine fibers, serve as a *chunking mechanism* for cortical processing. The amplitude, frequency, and phase of the excitatory and inhibitory fluctuations in ephatic potentials of slow-wave states will interfere, at the level of the junctional microstructure, positively or negatively with the coupling coefficients of individual neural elements, and affect the type of processing that is occurring (see Figure 13). The particular amplitude, frequency, and phase properties of the slow wave are referred to as window length for the input and processing of a particular chunk of information. The slower-frequency (8–12 HZ) higher-amplitude *alpha wave* would provide a relatively large window length, allowing for a more coherent and integrated form of processing, due to the relative largeness of the chunk and the consistency of the phase.

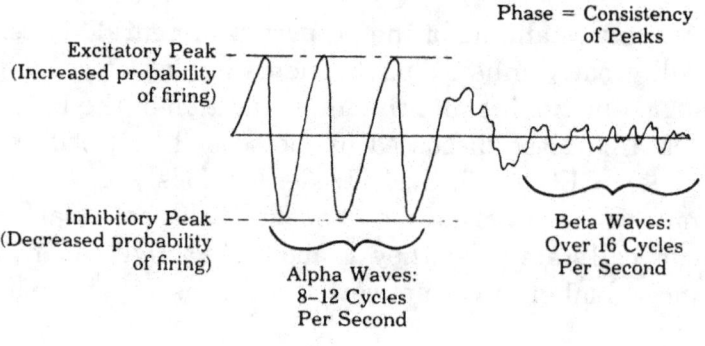

Figure 13.

The higher-frequency (16 HZ and over), lower-amplitude *beta wave* has much smaller chunks and much more arbitrary phase values. This, a typically more excited cortical state, would provide a smaller window length, allowing for the processing of quantitatively more information, although the processing may not be as complete or coherent as the slower frequency, more consistent phased alpha wave. The measurement of this type of state, more consistent phased over a particular area of the cortex, would indicate rapid processing, whereas the measurement of alpha waves would indicate slower, more efficient processing.

This would appear to fit in with the assumption, in many clinical and experimental circles, that the presence of EEG alpha indicates rest or nonactivity over that area of the cortex, while beta indicates cognitive effort or attention. (This was the assumption used in much of the hemispheric dominance research.)

This also seems to fit in with the writings of Callaway (1975) who, working largely with averaged evoked potentials (AEPs), cites the notion of *neural efficiency*. *Averaged evoked responses* are determined by adding together a number of recordings of neural electric activity evoked by the presentation of a particular stimulus. The parameters to be considered here are: 1) amplitude 2) frequencies 3) latency (the time between when the stimulus is presented and the electrical response appears) and 4) variability (changes in the evoked response from trial to trial). The neural-efficiency hypothesis derived from evoked response experiments done with individuals of widely varying IQs. These experiments tended to show largely differing AEPs of individuals with higher and lower IQs in response to flashes of light (see Figure 14), particularly in frequency and latency of response. The hypothesis was that individuals of higher IQ simply had more efficient neural structures.

It is interesting to note, however, that such a correlation did not hold for the presentation of auditory stimuli, nor in terms of an actual task carried out in response to the stimuli. Interpreting this data in terms of the notions of representational systems and strategies, one might hypothesize that the data presented in Figure 14 is actually measuring neural efficiency for visual information. The individual with the highest IQ would be considered to have a more highly developed visual system and visual strategies than the others. It would be interesting to see whether there would be a reversal of this pattern of efficiency if the AEPs had been taken in response to the presentation of a kinesthetic stimulus.

This type of analysis insinuates, of course, that an IQ test would require primarily visual processing strategies, and that the individual of the lowest IQ would probably tend to perform better at tactile, manual and mechanical tasks (which tends to be, incidentally, what individuals of lower IQ end up doing).

VIII. The Holographic Brain

Perhaps the most elegant model, with which to interpret neurological processing states, is the *holographic* or *holonomic* model proposed by Karl Pribram (1974, 1976). Pribram likens the interference patterns created by the interactions between the state of the junctional microstructure and syn-

EEG and Representational Systems 43

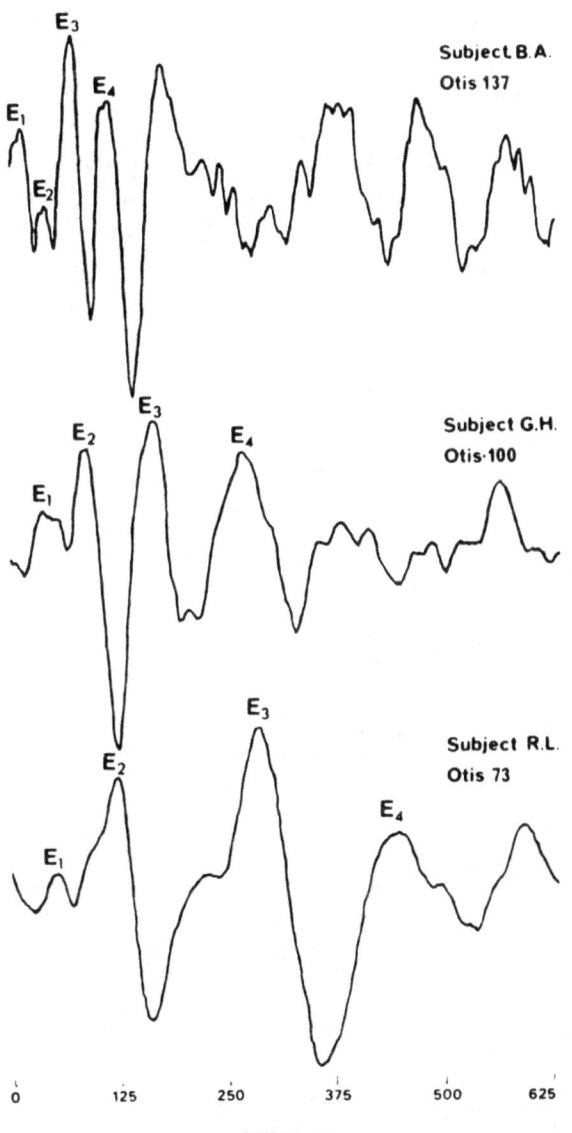

Figure 14. AEPs to 400 photic stimuli for three subjects of widely differing intelligence. Note progressive increase in latency with decreasing Otis I.Q. test scores.

aptic coupling coefficients (and their subsequent effects on membrane firing properties) to the interference processes between light waves involved in the establishment of an optical hologram (Figure 15).

Pribram proposes that any complex system composed of input-transform-output planes, in which inputs interact to form recordable interference patterns at the transform plane, is capable of producing direct transforms, at the output plane, of the

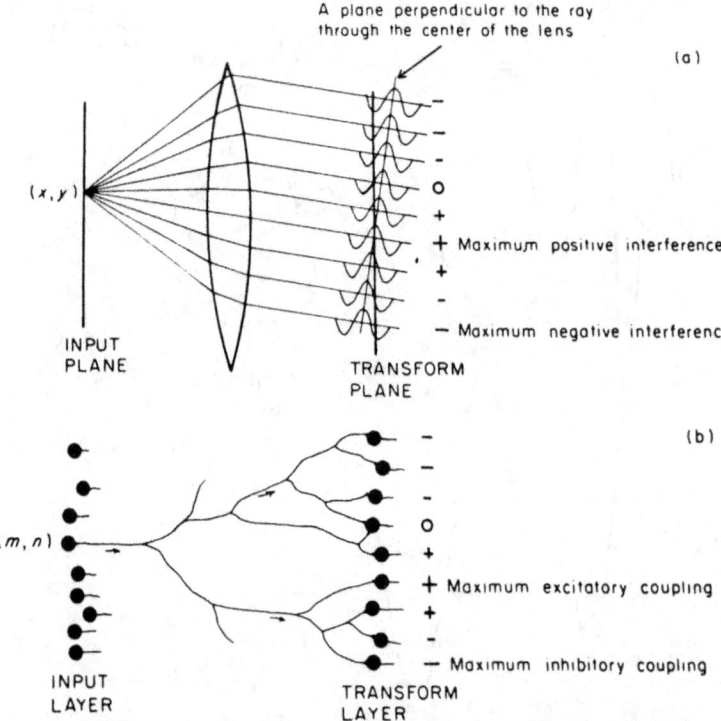

Figure 15. Correspondence between optical and neural systems. The couplings in the neural system may be considered with respect to some positive baseline. Thus an 'inhibitory' coupling is in reality just the decrease in excitation below baseline level, and still could be excitatory in physiological terms.

input signal when any part of the original signal is reintroduced. The strength of the output will be dependent on the amount and amplitude of the original input state (the total input effects on the transform plane) that is reintroduced. (One might postulate immediately that a coherent ephatic cortical state, like alpha waves, might provide the strongest recall stimulus.)

The recording of these interference patterns is thought to take place within the neural membrane. Pribram proposes that the interactions at the synaptic junctional microstructure:

> ... produce significant electric fields or voltage gradient components that are parallel to the surface of the membrane. We propose that these transient horizontal components of electric field trigger structural (for example, conformational) changes in the membrane that outlast these horizontal fields. The induced structural changes in the membrane which in themselves may be reversable could then set off further biochemical processes leading to long-lasting ion permeability changes. (see Figure 16)
>
> When either synapse is activated again, these structural or permeability changes can cause the effects (i.e., postsynaptic potentials) of one synaptic input to diffract and mimic the effects of the other, as if the latter were present. Thus the activation of one synapse produces the effects of activating both. The contribution of any such pair of synaptic inputs is small, but when many identical effects throughout the microstructure of the synaptic domain are summed, the physiology of the network would be significantly affected. (p. 444)

This concept is an extension of the "beaten-path" notion of learning. It adds in, however, the possibility of single trial learning and state dependent learning.

We discussed earlier the possibility that the strongest and most lasting membrane potentials might take place at various critical periods in devel-

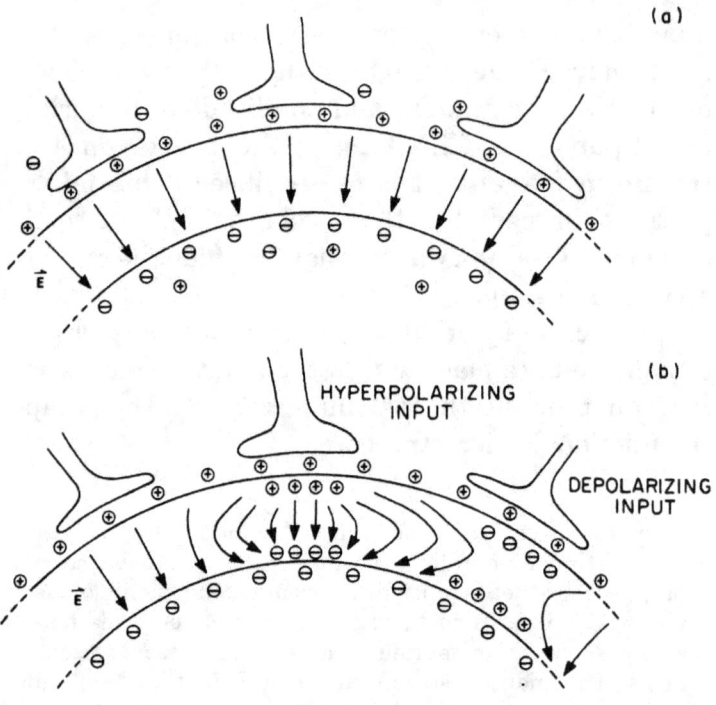

Figure 16. Membrane electrical fields. The upper figure (a) shows the field when undisturbed. The lower figure (b) shows the field when inputs arrive, demonstrating the consequent horizontal components of the field within the membrane. (From Richard Gauthier, 1972.)

opment. In terms of the neurological representation of strategies, one may consider that the development of a particular formal strategy is the result of this process taking place at lower level structures (the thalamus, for instance) in which membrane permeability changes tend to channel more or higher signal values for activity within a particular infrasystem.

Another important aspect of this model of neural functioning is that it predicts distinct forms of

EEG and Representational Systems

output, depending on the type of input. Pribram states:

> "For an arbitrary input pattern (that is, the encoded form of the sensory stimulus) (beta activity—R.D.), the output pattern is the cross-correlation of the input pattern represented by the local sensitivity values. In this case, the system gives a strong signal when the input pattern is similar to the stored pattern. This is the recognition process. In the optical system ... the recognition signal appears as a bright spot in the output plane of the system. By analogy, the recognition signal for a memory node is the rapid firing of a small group of cells in the output layer. The sensitivity pattern has a 'focusing' effect on the surrounding activity, and when the input pattern is similar to the stored pattern, the recognition information is gathered at that group of cells. The activity in such cells could easily be monitored by other networks and used for selecting the memory nodes from which to later recall information ...
>
> "By contrast, the output pattern has a completely different nature when the cells of the transform layer are excited in a uniform way. If the memory cells are uniformly excited, that is, by a single cell whose coupling coefficients are the same to all memory cells, the uniform activation is modified by the sensitivity values of the memory cells and the result is that each memory cell fires at a rate that is proportional to its sensitivity value. In this case, their pattern of depolarization is a reactivation of the stored pattern ... The output pattern is, in this case, a copy of the original pattern of departure activity that was input when the local sensitivity values were established. The output pattern is a recalled copy of the stored information." (p. 441)

The uniform input pattern here would of course be the stable amplitude and frequency values of the alpha rhythm.

Presence of beta activity measured over a particular area of the cortex, then, would indicate the recognition and channeling of inputs to be processed by focusing activity into signal nodes of higher intensity. These channeling effects will undoubtedly contribute to the establishment of recurrent strategies and of a most highly valued representational system that was discussed before. Presence of the slower and more coherent alpha waves would indicate memory search processes.

Such an interpretation might cast some new light on some of the hemispheric lateralization and specialization research. Here, the lateralized presence of alpha would not simply indicate that one hemisphere was working and the other idling. But rather, that each was performing important, but qualitatively different, functions.

IX. The Neurology of Learning

Pribram's theory seems to fit in extremely well with the research of E. Roy John (1975), who has done work with electrodes implanted in the brains of cats. John notes that the averaged evoked response to a particular conditioning stimulus is generated by stimulus-bound neurons (class I). Electrical patterns picked up by both surface EEG and deep probes show an invariant average response to a given stimulus, no matter what behavioral response ensues.

He points out, however, that during conditioning or learning the anatomical distribution of evoked responses to a stimulus becomes more widespread. (This is most likely the result of the increase in input signal values caused by the recognition process described by Pribram.) And, that while responses continue to be displayed by sensory specific structures of the modality of the signal, new responses appear in regions that showed little or no response to the conditioning stimulus before training. Furthermore, John notes that, when an expected event does not

EEG and Representational Systems

occur, these new cerebral potentials appear at a latency similar to that of potentials usually evoked by the expected stimulus. He calls this class of activity *readout* or *emitted potentials.*

John writes (1975):

> The critical event, learning, is envisaged as the establishment of ... representational systems (not the same meaning as in NLP—R.D.) of large numbers of neurons in different parts of the brain, whose activity has been affected in a coordinated way by the spatiotemporal characteristics of the stimuli present during a learning experience. The coherent pattern of discharge of neurons in these regions spreads to numerous other regions of the brain. Sustained transactions of activity between participating cells permit rapid interaction among all regions affected by the incoming sequence of stimuli as well as the subsequent spread. This initiates the development of a *common mode of activity,* a temporal pattern which is coherent across various regions specific for that stimulus complex. As this common mode of activity is sustained, certain changes are presumed to take place ... (that) increase the probability of recurrence of that coherent pattern in the network.
>
> "... The same ensemble can represent many different items, each with a different coherent pattern ... new responses are based upon the establishment of new temporal patterns of ensemble activity, rather than on the elaboration of new pathways or connections. Learning increases the probability that particular temporal patterns will occur in ensembles of neurons. By this process, the representational system acquires the capability of releasing the specified common mode of activity as a whole, if some significant portion of the system enters the appropriate mode.

John calls potentials initiated by *stimulus bound* activity *exogenous activity,* and those generated through *internal* means *endogenous.* John found, through the use of implanted electrodes, that readout processes arise from hierarchically lower brain structures (the cortico-reticular region) and propagate to involve other brain areas in a systematic sequence, through cortico-reticular feedback and synaptic gates.

By first averaging and recording the evoked response to a conditioning stimulus, before learning had taken place, John found that he could subtract the endogenously evoked wave shape, via a computer, from evoked response observed after learning to produce an accurate electrical facsimile of the purely endogenous activity. In a dramatic demonstration, he then took a number of cats that had been overconditioned to carry out certain responses to particular frequencies of auditory and visual stimulation, and played back their recorded readout potentials through electrodes implanted in the reticular formation of their brains. The result was the conditioned response, even though the appropriate sensory cue had not been given.

The establishment of a particular strategy or pattern of behavior, then, comes from the inputting of a particular temporally coherent pattern of ephatic activity into the neural metasystem. When the pattern has consolidated, it is initiated by structurally lower brain structures, in particular, the reticular region.

The process of learning, then, involves two major parts: 1) a focused input signal to program neural circuits at the synaptic level; and 2) once the program is established, a steady coherent input is required to reinitiate stable activity in the program circuitry. Behaviors that are well learned (that is, overconditioned and habitual) are correlated directly with the establishment of coherent slow potentials.

X. Neurophysiology of Consciousness

The reader has probably already noted that all of the steps and representations of a particular strategy or pattern of behavior need not be conscious to the individual, who has incorporated it, to be effective. In fact, generally, just the opposite is true—the more habitual and less conscious a behavior becomes, the more we can be guaranteed that we will have completely incorporated it. The process of learning to drive a car is the process of becoming less and less aware of all of the individual steps (pressure on the gas pedal, distance judgment, and so on) so that the behavior may be established as a single unit. Similarly, there are probably very few of us who need to be conscious of all of the individual cognitive steps we must go through to speak our native language (although at one time, as children, we did have to go through this process). Habituated and unconscious learning has been correlated directly with the establishment of slow wave coherent activity. People, for instance, are typically less conscious of internal cognitive processing during an alpha state.

Consciousness is, however, very important for the initial spread and establishment of a particular strategy. In the Neuro-Linguistic Programming model, consciousness is considered to be a function of the signal values of activity within the various representational systems—the highest signal being the most conscious. It is important to keep in mind that consciousness is a transform of how, and how much, a particular representational system (at the neurological level) is being used, rather than being an entity that initiates activity itself. (To say that the property of consciousness or awareness controlled or affected behavior would be like saying that the properties of wetness or iciness controlled or affected the structural combinations of H_2O molecules.)

Consciousness (high signal) is necessary at the time when a strategy is being established (in order to provide a strong enough input to change synaptic coupling coefficients sufficiently), but once the program is established, the high signal (determined by latency, frequency, and focusing) is replaced by the coherence and phase of the habituated slow potential.

XI. Accessing Cues: Tuning in Brain Functions

It has been estimated that human beings process tens of thousands of bits of information each second. Yet we know that, in order to respond in an organized fashion to this information, our brains must filter all this information considerably. A final issue, of primary importance, is how an individual is able to tune internal states to provide for the cycling of signal values required for the carrying out of a cognitive strategy. Bandler and Grinder (1976) call a behavior used to cause an alteration in an individual's mental/physiological state as a means to tune into modality-specific activity an *accessing cue*. State-accessing processes range from *breathing rates* and position (chest or abdomen), to the degree of *muscle tonus* in various muscle groups, *body temperature*, metabolism, hormonal changes, *ocular-motor activity* (eye-movements), *heart rate* and *blood pressure*, even certain types of *vocalizations*. Practically any process will contribute, in some way, to state fluctuations, in that any occurrence in one part of a cybernetic system (such as a human being) will necessarily affect all other parts of that system in some way. It is on the learning of accessing sequences, then, that much of the consolidation of our particular strategies will rely. It should be noted that a particular accessing cue is not thought to "cause" an image, sound or feeling, but is rather considered to function similarly to the tuning knob on a radio. Many radio stations are constantly broadcasting (just as our bodies are constantly bombarded by sensory information), but by altering the internal works of the radio the signal value of one station may be tuned with the minimum amount of interference from the others.

An important result of this is that through attention to observable accessing cues, one may also get information about the representational form of a particular individual's strategies. Because all behavior, microscopic or macroscopic, is a transform of internal neurological processes, it will carry information about those processes. All behavior then is in some way a communication about the internal neural organization of an individual. By simply observing lateral eye movements or listening to an individual's choice of words, one can gain a great deal of information about the way that individual organizes his/her experience. EEG activity is another such transform.

Up to this point, accessing cue patterns have been established primarily through clinical observation. Indeed, many accessing patterns are so transient and complex that it is difficult to quantify them experimentally.

Because of their importance to the research work

to follow this text, I will identify four major accessing patterns that have been clinically verified, as a means to provide the reader some tools to identify and index some of the representational parameters of cognitive strategies:

1. *Lateral Eye Movements* (see Figure 17)

a) Up and Left—accesses non-dominant hemisphere visualization (for right-handed people), that is, remembered or eidetic imagery.

b) Up and Right—accesses dominant hemisphere visualization (in righthanders), that is, constructed imagery and manipulation of visual images.

c) Level and to the Left—accesses non-dominant hemisphere processing of sounds and words (in righthanders), that is, remembered sounds and "tape loops" (overconditioned phrases such as nursery rhymes and clichés).

d) Level and to the Right—indicates dominant hemisphere auditory processing (for righthanders), that is, construction of things to say, translation, and manipulation of internally generated sounds.

e) Down and Left—accesses internal dialogue (in righthanders), that is, talking to and with oneself, internally.

f) Down and Right—accesses (in righthanders) both emotional and tactile feelings.

2. *Predicates*—Grinder and Bandler have shown how sensory-specific words, which people use, will often be an accurate indicator of internal processes.

a) *Visual*—I *see* what you are saying; That *looks* good; That idea isn't *clear* to me; I went *blank;* Cast

EEG and Representational Systems

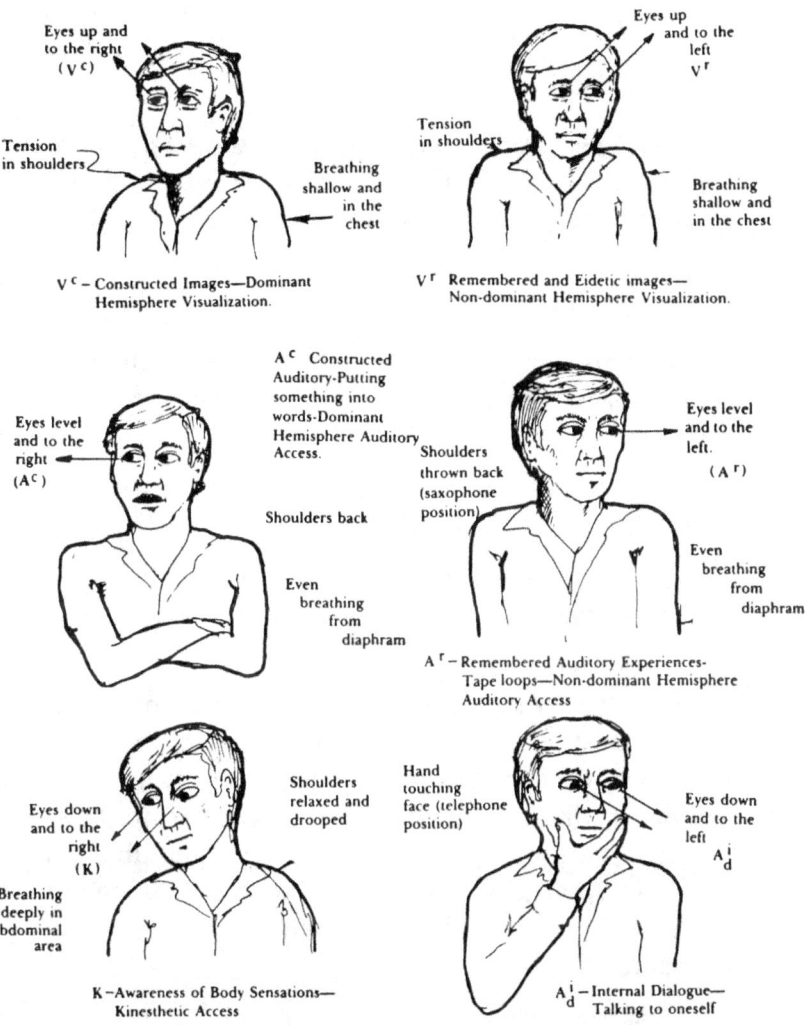

Figure 17. Accessing cues for typically wired right handed person.

some *light* on the subject; Get a new *perspective;* An *(insightful, enlightening, colorful)* example.

b) *Feelings*—It doesn't *feel* right; She's so *cold;* (Get a *handle* on, *grasp*) that idea; Get in *touch* with that; He's *insensitive;* a *solid* understanding; I have a *feeling* you're right.

c) *Auditory*—I *hear* you; That *(rings a bell, sounds good); Listen* to that; Something *tells* me that I shouldn't; It just suddenly *clicked;* Get in *tune* with yourself.

3. *Tonal and Tempo Changes*

a) *Visual*—High-pitched nasal and/or strained tonality. Speaks with quick bursts of words and generally fast tempo.

b) *Feelings*—Low, deep, breathy voice with slow tempo, and pauses between words.

c) *Auditory*—Clear resonant tonality with an even rhythmic tempo.

4. *Body Types*—Body types are more indicative of an individual's most highly valued representational system as they reflect representational strategies that have occurred over long periods of time. They are determined by interaction of other state determiners such as muscle tension and metabolism. They, incidentally, tend to correspond, somewhat, with the constitutional typology proposed by Sheldon (1942).

a) *Visual*—thin tense body type (ectomorph).

b) *Tactile kinesthetic*—active muscular body type (mezomorph).

c) *Visceral kinesthetic*—heavier, fuller, softer body (endomorph).

d) *Auditory*—tend to be in between visual and auditory body type although primarily ectomorphic;

tend to have a posture involving shoulders thrown back and head forward.

Given this background, we can now move on to a discussion of the research project that was an effort to correlate EEG readings with the concept of representational systems and strategies.

THE EXPERIMENT

PURPOSE:
The purpose of this research was to attempt to correlate lateral eye movements with sensory-specific, internal activity within an individual's ongoing subjective experience, and the concomitant electrical brain activity. Specifically, we attempted to index subjectively experienced and reported internal images, voices and sounds, emotional responses, and body sensations with systematic and recurrent eye positions and changes in electroencephalogram (EEG) recordings.

This project was based on the clinical research of John Grinder and Richard Bandler (1976) that suggested that lateral eye movements will index internal representations in the following manner (hemispheres indicated are for right-handed people):

1. *Up and to the left*—indicates non-dominant hemisphere visualization, i.e., remembered or eidetic imagery;

2. *Up and to the right*—indicates dominant hemisphere visualization, i.e., constructed imagery and manipulation of visual images;

3. *Level and to the left*—indicates non-dominant hemisphere processing of sounds and words, i.e., remembered sounds and "tape loops" (overconditioned phrases such as nursery rhymes and clichés);

4. *Level and to the right*—indicates dominant hemisphere auditory processing, i.e., construction of things to say, translation, and manipulation of internally generated sounds;

5. *Down and to the left*—indicates internal dialogue, i.e., talking to and with oneself internally;

6. *Down and to the right*—indicates access of both emotional and tactile feelings;

NOTE: These patterns would be reversed from left to right for a completely left-handed individual.

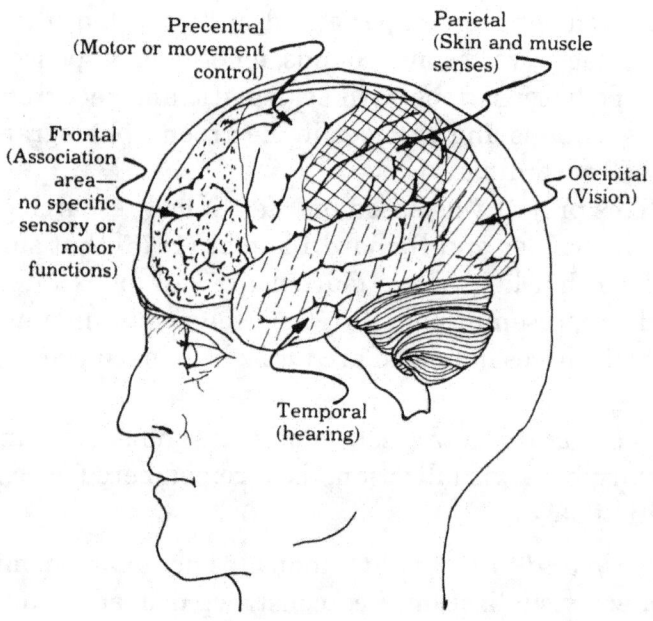

Figure 1.

Since it has been shown in studies using electric stimulation and the average evoked response (AER) that various areas of the human cortex seem to be responsible for processing sensory-specific activity, it was hoped that we could correlate changes in EEG activity in sensory-specific cortical areas with simultaneous eye movements. The major cortical subdivisions and their sensory associations are shown in Figure 1.

Equipment

The research project was conducted in the laboratory of Joe Kamiya, professor of psychophysiology at the Langley-Porter Institute, a division of the University of California Medical School in San Francisco. The major equipment components required for the research were:

1. A shielded screening room for the subjects;

2. Eleven (11) surface electrodes to record physiological electrical fluctuations—four electrodes for electro-ocular activity (EOG), the remainder to record cortical activity;

3. A Grass channel amplifier and polygraph to amplify and record four channels of EEG cortical electrodes;

4. A Beckman channel amplifier (wired to the Grass polygraph recorder and pens) to amplify and record eye movements;

5. An on-line PDP-11 computer programmed to receive, store, and analyze electrical wave shapes picked up by the EEG surface electrodes.

Experimental Setup

The equipment was set up to function in the following manner:

1. The eye electrodes were placed as two pairs,

V_1/V_2 and H_1/H_2, to pick up vertical (V_1 and V_2) and horizontal (H_1 and H_2) eye movements. These movements were plotted on a six point matrix (up, level, down and right, left). H_1 was placed to the left of the left eye, H_2 to the right of the right eye. V_1 was placed on the eyebrow above the left eye, V_2 just below the left eye (only one eye was needed to record vertical movements since both eyes move in unison). Because there is a standing electrical potential between the retina and cornea of the eye, the electrodes picked up the changes in the amplitude of the standing voltage as the eye moved closer to a particular electrode. The electrodes were wired to the polygraph so that this change in voltage caused a pen to move up or down, depending on which electrode the eye moved closer to. The voltage going to the polygraph varied in response to the difference in potential between V_1 & V_2 and H_1 & H_2 (see Figures 2 and 3).

2. The surface cortical electrodes were placed in symmetrical pairs over the occipital and central lobes of the right and left cerebral cortex. (In a few cases the second pair was placed over temporal lobes.) The activity from each cortical placement was measured against a neutral source—the left and right earlobes (A_1 and A_2, respectively). The activity from the left occipital lobe (O_1) and left central placement (C_1) was referenced to the left earlobe (A_1). Activity from the right occipital cortex (O_2) and right central cortex (C_2) were measured against the right earlobe (A_2). (Temporal placements, when used, were labeled T_1 and T_2 in the same manner as the occipital and central.) A frontal placement, F, was used as a ground for the whole system of electrodes.

Using this configuration (see Figure 4), the EEG

The Experiment

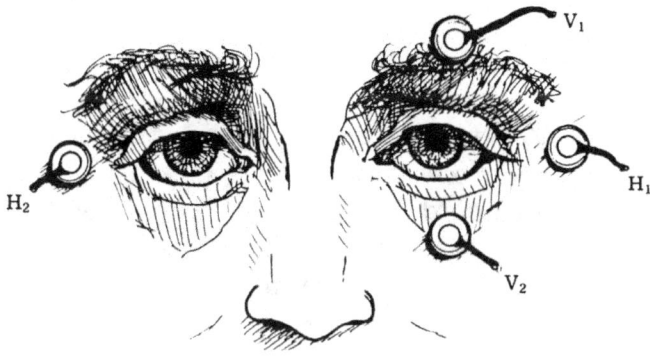

Figure 2. EOG eye electrode placement.

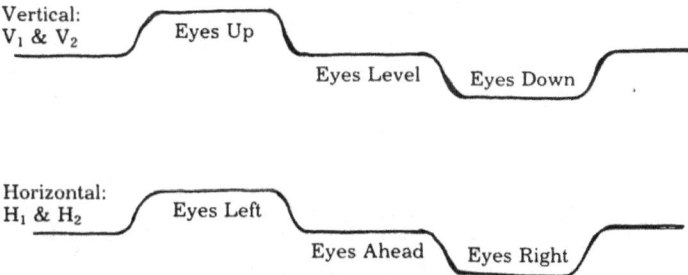

Figure 3. Polygraph pen recording.

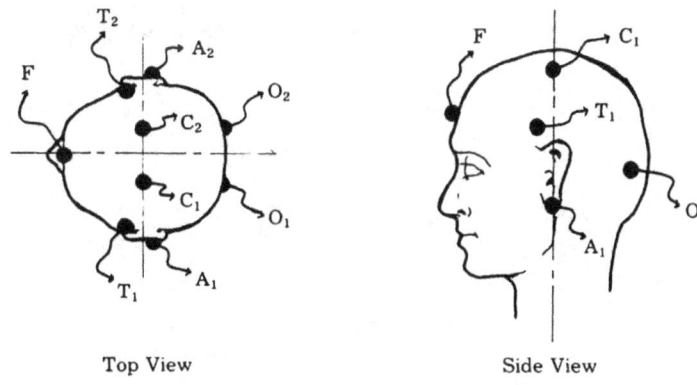

Figure 4. Surface electrode placement.

picked up electrical transforms of visual (O_1 and O_2) and kinesthetic/motor (C_1 and C_2) responses (auditory responses, T_1 and T_2, were also recorded on occasion) over both cerebral hemispheres. This procedure gave a record of differences between hemispheric activity of the same cortical origin as well as a record of individual cortical activity.

The recording pens for the EOG and the EEG were run on the same strip chart to give a running record of simultaneous lateral eye movements and EEG activity. See Figure 5.

3. The computer was programmed to take in, store, and analyze data at intervals (epochs) of 16 seconds.

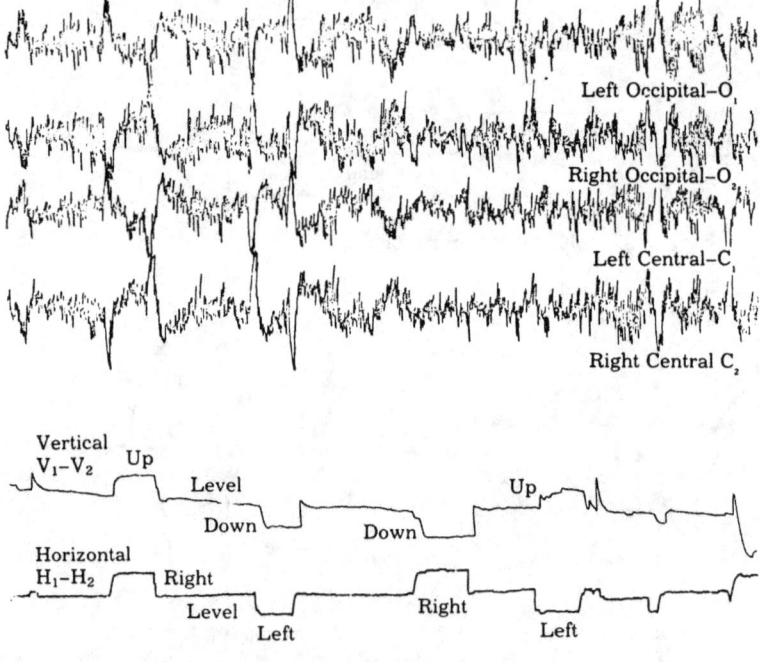

Figure 5. Polygraph recording of four channels of EEG and two channels of EOG operating simultaneously.

The Experiment

Each 16-second epoch was made up of eight, smaller, 2-second intervals for finer details. The computer data was compiled in three parts:

A) A listing of the total band amplitude (all waveforms between .5 and 60 Hz) for each channel of EEG at two-second intervals and a left/right index displaying the relative differences in overall activity for the left and right hemispheres for both of the cortical placements. The percentage of difference was computed by subtracting the average total activity in the right hemisphere for a two-second interval from the average total activity in the left hemisphere and dividing by their sum: $\frac{L - R}{L + R}$ —in this case $\frac{O_1 - O_2}{O_1 + O_2}$ and $\frac{C_1 - C_2}{C_1 + C_2}$.

B) An average amplitude and laterality score for electrical wave frequencies in or near the alpha band (8–13 Hz). The lateralized alpha waves were measured by the percentage of difference, i.e., the percentage of total alpha in the left hemisphere minus the percentage of total alpha in the right hemisphere, divided by their sum: $\frac{\alpha L - \alpha R}{\alpha L + \alpha R}$

The amount of alpha in each hemisphere was calculated by integrating voltage amplitudes (filtered for frequencies between .5 and 14.5 Hz) from each cortical channel over the 16-second epoch. The numbers were inserted into the expression for percent of difference to yield a laterality score for both cortices (occipital and central) in each epoch.

The alpha band of the EEG was chosen as a dependent variable in this study because it seems to change while cognitive processing is taking place. The alpha band has been associated with a relaxed and resting condition of minimal cognitive activity.

The alpha rhythm characteristically disappears with any external stimulus or with intense cognitive effort. Increased alpha is correlated with increased relaxation and decreased activity. In some recent experiments that measured the amount of alpha separately from each hemisphere, the ratio of lateralization was compared as a function of task. The exact comparison figure varied from one laboratory to another, as does the method for calculating alpha. The findings, however, indicated that the proportion of alpha in the right hemisphere to the total amount decreased in visuo-spatial tasks and increased in linguistic and mathematical tasks. That is, there was alpha suppression, relative to the total amount of alpha, in the hemisphere that was processing the information.

To examine possible correlations between the left and right lateral eye movements and the tasks they seem to index, we decided to record information about the lateralization of the alpha band.

C) Fast Fourier Transforms (FFT) were used to break down frequency and amplitude data into half-Hertz chunks. That is, amplitudes were computed for each frequency increase of ½ Hertz for frequencies between .5 and 20 Hz. This gave amplitude distribution information for the alpha band. See Figure 6.

Procedure

The experiment proceeded in the following manner:

1. After the subject was connected to the electrodes, he/she was asked to sit in the experiment room and "think of nothing in particular." Baseline

**** EPOCH 2 SN 50: 123.0 = TIME OF AWAKENING

SAMPLE TIMES: VR - REMEMBERED VISUAL TASKS - EPOCH = 16 SECONDS TOTAL TIME

122.0 120.0 118.0 116.0 114.0 112.0 110.0 108.0 (EARLIEST)

L-R INDEX, L, R — TOTAL BAND AMPLITUDE - TAKEN FOR EACH CHANNEL AT 2 SECOND INTERVALS

	O_1-O_2	O_1	O_2	C_1-C_2	C_1	C_2
MOST RECENT INTERVAL 5	0.067	35.59	31.10	-0.004	21.13	21.30
	-0.052	19.98	22.20	-0.049	21.98	24.25
	-0.113	20.69	25.98	-0.013	20.32	20.86
	0.048	22.79	20.69	-0.025	21.54	22.67
	0.022	21.71	20.74	0.022	20.96	20.03
	-0.086	17.86	21.25	0.015	19.45	18.86
EARLIEST	-0.111	15.62	19.52	-0.025	16.59	17.45
INTERVAL 2	0.075	19.15	22.30	-0.027	18.37	19.42
	-.038	22	23	-.013	20	20.5

O_1-O_2 is computed by taking $L-R/L+R$ or: O_1-O_2/O_1+O_2

LEFT-RIGHT INDICES, L, & R, BY BINS — ALPHA BAND TOTALS FOR 16 SECONDS

	.5-1.5 Hz	2-3.5 Hz	4-5.5	6-7.5	8-9.5	10-11.5	12-14.5 Hz
	-0.079	0.070	-0.044	-0.070	-0.044	-0.168	-0.096
O_1	134.73	73.64	50.24	68.18	183.50	63.58	32.55
O_2	158.04	63.88	54.90	78.54	200.70	89.34	39.52
	-0.036	0.024	0.004	-0.019	-0.042	0.066	-0.004
C_1	156.86	85.14	64.46	75.87	120.96	73.40	31.45
C_2	168.79	81.01	63.85	78.91	131.70	64.22	31.74

→ O_1-O_2/O_1+O_2 for α band; → C_1-C_2/C_1+C_2

FFT DATA

AMPLITUDE SPECTRUM FOR FREQUENCIES BETWEEN 0 & 19.5 HZ

		.5	1	1.5	2	
1	0 66.40	193.58	126.81	83.81	102.71	
	2.5 68.65	58.37	64.82	47.86	47.57	4.5
	5 46.90	58.54	63.45	78.18	56.12	7
	7.5 74.98	109.07	234.24	280.20	110.51	9.5
	10 108.69	63.99	52.84	28.78	42.36	12
	12.5 26.28	28.25	28.49	28.84	41.07	14.5
	15 32.74	31.85	33.83	40.02	52.24	17
	17.5 39.48	47.75	23.43	42.36	23.33	19.5 O_1
		18	18.5	19		
2	0 61.73	242.63	133.25	98.24	87.60	2
	66.81	60.01	41.08	46.97	50.24	
	53.73	68.68	50.23	75.96	81.08	
	106.88	112.07	287.41	244.07	159.25	
	169.67	94.88	41.42	51.37	47.12	
	37.08	34.58	51.72	30.33	36.30	
	29.30	35.44	41.11	48.24	43.51	
	17.5 44.05	42.74	56.51	55.68	30.96	19.5 O_2
3	0 65.77	226.46	114.62	129.49	91.71	2
	92.01	74.39	82.44	65.94	67.55	
	50.16	74.20	70.65	54.62	65.46	
	112.75	104.39	145.86	147.01	86.60	
	106.67	78.82	62.84	45.27	31.05	
	28.91	34.86	24.51	28.49	40.87	
	43.80	33.51	31.16	27.03	24.90	
	17.5 24.46	40.19	19.22	38.34	34.07	19.5 C_1
4	0 71.19	255.44	128.41	122.51	88.54	2
	79.26	71.74	84.51	68.73	54.88	
	52.97	78.81	65.90	58.48	84.65	
	106.59	111.77	160.94	143.40	110.70	
	81.10	73.50	54.92	47.36	35.90	
	29.93	39.66	23.12	24.03	37.79	
	35.12	29.09	30.59	24.01	23.20	
	17.5 21.58	41.07	26.26	41.92	34.19	19.5 C_2

Figure 6. Sample of computer printout for a single epoch involving remembered visual images.

waves, to be used as a control, were then taken for two epochs.

2. The subject was then asked to close his/her eyes and again "think of nothing in particular," and two more baseline epochs were taken.

3. The subject was then asked a series of questions in eight groupings. Each grouping of questions appealed to a particular type of cognitive processing—visual, auditory, kinesthetic, and emotional (visceral feelings). Each was also geared to either memory (non-dominant hemisphere processing) or construction (dominant hemisphere processing). The subject was instructed not to answer the questions verbally, but to simply raise a finger when they had secured the answer, to avoid the possibility of any artifacts on the EEG that can result from talking.

A 16-second epoch was taken during each grouping to store and record electrical data occurring as the individual retrieved the cognitive information asked for by the questions. The questions were grouped as follows:

1. *Visual Remembered:* Think of the color of your car. What kind of pattern is on your bedspread? Think of the last time you saw someone running. Who were the first five people you saw this morning?

2. *Visual Construction:* Imagine an outline of yourself as you might look from six feet above us and see it turning into a city skyline. Can you imagine the top half of a toy dog on the bottom half of a green hippopotamus?

3. *Auditory Remembered:* Can you think of one of your favorite songs? Think of the sound of clapping. How does your car's engine sound?

4. *Auditory Constructed:* Imagine the sound of a

train's whistle turning into the sound of pages turning. Can you hear the sound of a saxophone and the sound of your mother's voice at the same time?

5. *Kinesthetic (Tactile) Remembered:* When was the last time you felt really wet? Imagine the feelings of snow in your hands. What does a pine cone feel like? When was the last time you touched a hot cooking utensil?

6. *Tactile Construction:* Imagine the feelings of stickiness turning into the feelings of sand shifting between your fingers. Imagine the feelings of dog's fur turning into the feelings of soft butter.

7. *Visceral/Emotional Remembered:* Can you think of a time you felt satisfied about something you completed? Think of what it feels like to be exhausted. When was the last time you felt impatient?

8. *Visceral Construction:* Imagine the feelings of frustration turning into the feeling of being really motivated to do something. Imagine the feeling of being bored turning into feeling good about feeling bored.

The experimenter was in the room with the subject while he was asking the subject to perform these cognitive tasks. The room was slightly illuminated and the subject was asked to keep his/her eyes open during the questioning. This was done to encourage some eye contact with the experimenter, which increases the probability of lateral eye movements; the subject must break away from external contact with the experimenter and "here and now" to access and retrieve information that is not in his/her immediate sensory environment.

It should be mentioned that these lateral eye

movements are not thought to actually "make" a particular cognitive process occur, but rather serve as triggers or accessing cues. A person, for instance, can learn to look up and to the left and not visualize; although often an individual won't be able to visualize until they do look in that direction. Although the mechanism is not understood, there does seem to be some innate neurological connection between these lateral eye movements and cognitive processes, because they have been observed to index the same states in individuals of different cultures and in individuals who have been blind from birth.

Figure 7 shows a sample of the strip-chart recording taken during a question-grouping epoch involving visual recall as a cognitive task.

4. The subject was then asked again to "rest comfortably and think of nothing in particular." Two more baseline readings were taken with eyes open and closed.

Results

A total of 25 subjects were run for the experiment. As is typical with an experimental situation, there were a number of technical difficulties with the equipment. The results, however, were obtained in spite of these problems, and their implications, I believe, will prove to be interesting and exciting.

1. It became obvious after the running of a few subjects that the lateral eye movements did not index any particular brain wave shape that was invariant across subjects. Rather, the particular wave shape accompanying an eye movement was relative to the individual's baseline control wave forms, which var-

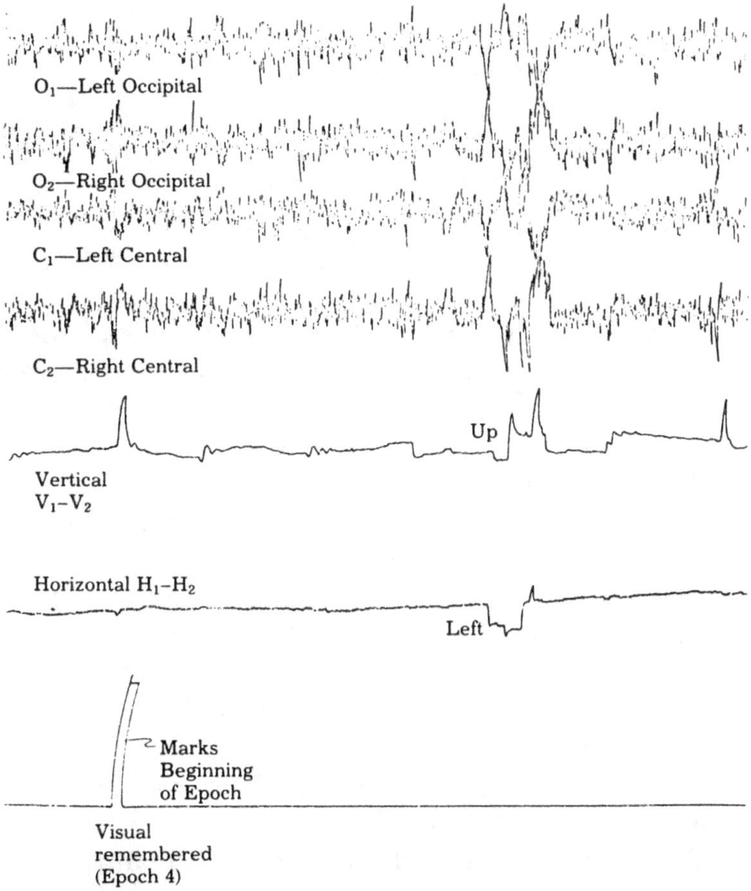

The Experiment

Figure 7. Sample strip chart reading taken during remembered visual question grouping.

ied significantly between subjects and groups of subjects.

When paired with the grouping of cognitive tasks and subjective reports of internal experience, however, there was an uncanny parallel between the position of the lateral eye movement and the type of cognitive processing. It is unfortunate that preparations for recording subjective reports were not included in this study. (I have used videotapes in a subsequent project. See Appendix A.)

The eye electrodes proved to be less satisfactory for plotting and recording eye movements than videotape or some other method (reflective contact lenses, for instance). Interference and artifact in the eye movement recordings made interpretation difficult at times.

Subjective reports proved to be of extreme importance in determining the particulars of cognitive processing occurring for a particular individual. An individual's report of their subjective experience often cleared up seeming discrepancies in observed eye movements. It became evident that more consideration to individual cognitive strategies was required. For instance, on some of the more complex tasks, individuals went through a sequence of cognitive representations and activity to arrive at the answer, and so displayed a more complex series of eye movements. Some individuals had to access the feelings of being in their car and watch their hand turning the ignition key before they could hear the sound of the motor. Some needed to physically manipulate (internally) the toy dog and hippopotamus before they could see them on top of one another.

Some individuals consistently needed to see the object or situation before they could hear it, feel it, or have feelings about it. Others needed to repeat the question to themselves or talk about it to themselves before they could retrieve or imagine any other sensory information. Still others needed to put themselves in situations kinesthetically before being able to access the information in question; while others lead with emotional and visceral attachments or judgments about the objects or situations, initially.

An unexpected and fascinating outcome of this consideration of cognitive strategies was the profound correlation between the most typical, or primary, cognitive mode (that is, auditory, visual, kinesthetic, or visceral) that an individual utilized and their baseline brain waves! Grinder and Bandler (1975, 1976) have identified a number of personality and behavioral traits that seem to be associated with the tendency of individuals to use one cognitive mode more than others: some people being more visually oriented; some more feeling oriented; others are auditorily prone.

In this study, primary cognitive mode was identified by the lateral eye position accessed most often and the representational modality cited most often by the subjects' verbal descriptions of their internal experience. The correlations were as follows:

Primary Modality	Control Wave Description
Visceral	High amplitude alpha content when eyes are both open and closed. (Fig. 8)
Visual	Low amplitude beta when eyes are open; spindles of alpha of intermediate amplitude when eyes are closed. (Fig. 9)
Kinesthetic	Low amplitude beta when eyes are open or closed. (Fig. 10)
Auditory	Higher amplitude beta with intermittent alpha when eyes are open or closed. (Fig. 11)

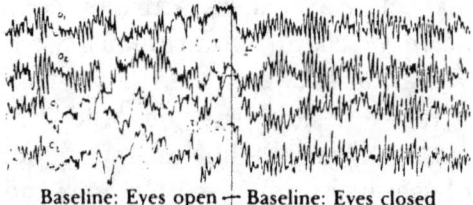

Baseline: Eyes open — Baseline: Eyes closed

Figure 8. Baseline EEG for individual of primarily visceral cognitive strategies.

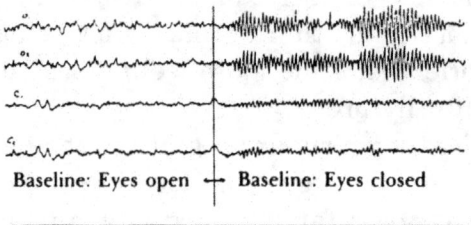

Baseline: Eyes open — Baseline: Eyes closed

Figure 9. Baseline EEG for individual of primarily visual cognitive strategies.

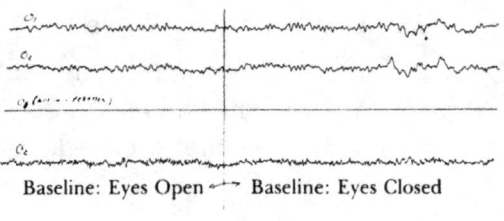

Baseline: Eyes Open — Baseline: Eyes Closed

Figure 10. Baseline EEG for individual of primarily kinesthetic cognitive strategies.

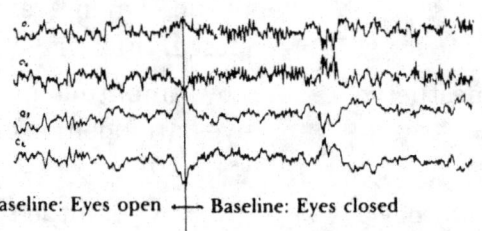

Baseline: Eyes open — Baseline: Eyes closed

Figure 11. Baseline EEG for individual of primarily auditory cognitive strategies.

The Experiment

This phenomenon is being studied further in a more appropriately designed experiment. (See Appendix A.)

It should be noted that these patterns show up best in the occipital (visual) area. It would seem to fit in with the notion that alpha activity indicates a state of rest or relaxation that individuals of primarily visual cognitive strategies would have beta activity over the visual cortex, while those relying on visceral information primarily would have predominantly alpha activity over the visual cortex. However, viscerally oriented people do use visualization, even though they often show co-occurring alpha waves.

Given my observation of patterns of alpha waves in the EEG recorded during this experiment, my tendency was to conclude that alpha activity did not index rest or inactivity, but rather activity of a different type. Karl Pribram (1974) discussed some of the differences between putting inputs of different types into a complex system such as the brain. He noted that inputting of an arbitrary fluctuating signal (such as beta activity) tends to cause recognition or channeling of the signal, depending on memory-changes in the system caused by past inputs. An even, steady signal tends to cause association or recall of information inputted by previous signals. In other words, activity such as beta would tend to cause or indicate information gathering, while alpha would tend to show memory.

Slow wave potentials, such as the alpha rhythm, have also been associated with the establishment of habitual, well-learned, over-conditioned behavior. Dancers and gymnasts, for instance, tend to have a high alpha count over the sensory motor area. Alpha, then, may indicate habitual patterns of activity. This could explain why a viscerally-oriented individual might have a lot of alpha over the visual area—most of their visual strategies may be preprogrammed and habitual. Similarly, when the visually-oriented individual closes his/her eyes (causing an immediate influx of alpha) he/she shuts off the primary information gathering device and must revert to pre-established programs.

Dumas and Morgan (1974) conducted an experiment comparing the EEG recordings of nine male artists with those of eight male engineers. The study was designed to test for any recordable asymmetries in the presence of alpha activity between the left and right occipital regions of subjects who had classically "dominant hemisphere" and "non-dominant hemisphere" occupations. They found that neither the artists nor the engineers "lived" in one hemisphere or the other in the sense that neither group showed any constant standing brain wave asymmetries. Rather, they found that brain wave activity lateralized (in terms of the presence or suppression of alpha and beta waves) in response to task specific activity (the subjects were given typical right and left hemisphere tasks such as face recall and spatial perception tests versus linguistic and addition tasks). The amount of lateralization (approximately 20%) was nearly the same for both groups; each group tending to lateralize to the appropriate hemisphere for the task.

They did, however, find a major difference in the standing baseline brain wave patterns of the two groups. They noted that the artists tended to have a high base line alpha content over the occipital

The Experiment

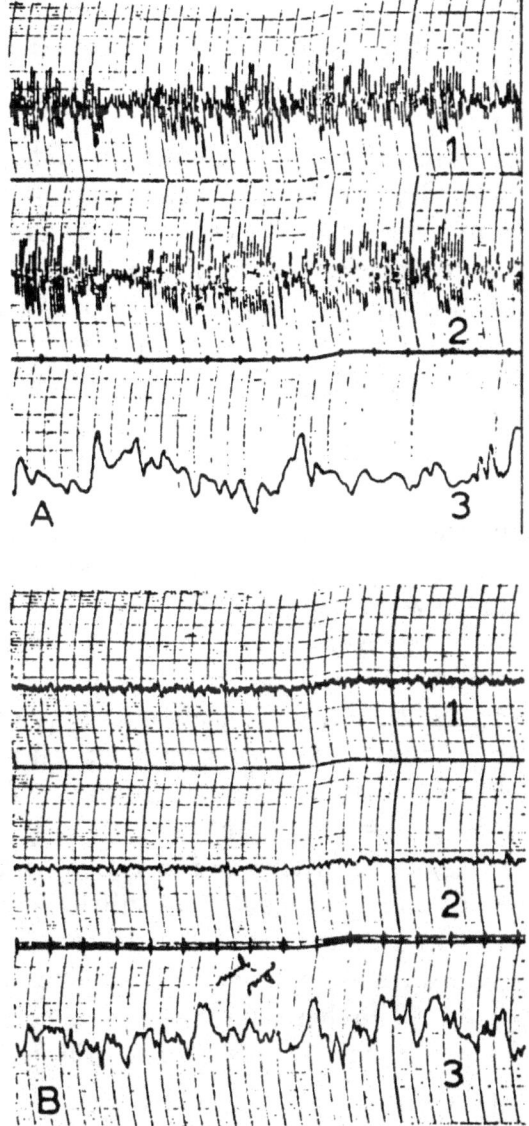

Figure 12. EEG record from an artist (A) and an engineer (B), showing EEG from the left hemisphere (line 1), the right hemisphere (line 2), and the instantaneous laterality (line 3). These two records were chosen to show the difference between the groups, but represent the extremes in amplitudes, not the average for each group.

area. On the other hand, the engineers tended to have much more beta activity in their base line readings (see Figure 12).

This seems to fit in quite well with the hypothesis I have proposed in that engineers are typically more visually oriented, while artists have more emotionally (viscerally) expressive strategies.

Another possibility is that slow wave activity (alpha) is associated with lower brain areas (such as the hypothalamus, thalamus, and other limbic structures) that are responsible for most visceral activity. Alpha activity may indicate the use of these structures.

It should be taken into consideration that skull thickness may cause amplitude and wave shape differences to show up on the EEG. In terms of my interpretation, however, I think that the patterns would show up even at different levels of amplitude.

2. In terms of the lateralization of activity during sensory-specific tasks, my recordings tended to confirm other tests which showed lateralization of brain activity during different cognitive tasks. There were some cases, however, where an individual (who was right-handed) would show a typical shifting of alpha to the right hemisphere for constructed tasks within auditory, visual, and kinesthetic modalities, but a shift in the opposite direction for visceral construction. This could make interesting material for a future study.

Conclusions

My conclusions are that lateral eye movements do indeed index modality-specific cognitive activity, but that the setup of this particular project was not the most efficient way to study them. More research needs to be done (and is being done) on the indexing of base line brain activity and what it indicates. Concurrent videotaping and recording of subjective responses and reports would have also helped out a great deal (I added this into a subsequent experiment—see Appendix A).

More attention to individual cognitive strategies and other behavioral indicators of such ongoing strategies would also assist in the interpretation of the EEG output. I pursued this also in subsequent research.

The epoch length of 16 seconds—from which the data for the computer was taken—was perhaps a less effective means of scoring the brain wave data that was to be paired with eye movements. A more immediate and detailed measure, such as recording evoked responses triggered by lateral eye movements, might have provided more specified results.

APPENDIX A

Purpose:

The purpose of this experiment was an attempt to verify the hypothesis made as a result of the information gathered in the experiment described earlier in this text. Specifically, the hypothesis that individuals of a particular primary cognitive mode (visual, auditory, visceral, or tactile kinesthetic) will have identifiable patterns of EEG baseline control waves as they sit with their eyes open and closed. The correlations to be verified consisted of the following observed patterns:

Primary Modality	*Control wave description*
Visceral	High-amplitude alpha content when eyes are both open and closed.
Visual	Low-amplitude beta activity when eyes are open; Spindles of alpha (of varying amplitudes) when eyes are closed.
Kinesthetic	Low-amplitude beta when eyes are open or closed.
Auditory	Higher-amplitude beta with some intermittent alpha when eyes are open or closed.

The patterns listed above were identified from activity taken primarily from the occipital area (visual

The Experiment

cortex). (See examples of these waves in Figures 8-11.)

Experimental Setup:

The setup of the experiment was fairly straightforward and consisted of the following procedures:

1. Fifty subjects were given a specially designed interview. The interview consisted of six questions designed to elicit primary representational strategies. (See sample questionnaire-scoresheet.) The questions involved such issues as importance, meaning, interest, learning abilities, truth, altered states of consciousness, and self-image. Each interview was then scored, using the scoresheet in Figure 13.

Individuals were scored in their responses to each question along the following two parameters: 1) position and sequence of lateral eye movements; and 2) types of predicates chosen during their response. The correlation between these behaviors and the sensory-specific activity (outlined, previously, in this text) were presupposed as an accurate index of cognitive activity.

Lateral eye movements were scored using a nine-point matrix (up-level-down, and right-middle-left). The positions and sequence of eye movements were then plotted on the matrix by putting digits, in numerical order, in the appropriate boxes and indicating the appropriate temporal steps of the strategy. Eye position at the time of any vocalization was circled, as it was postulated that this access would be most highly valued for that answer.

Any relevant or sensory-specific predicates were

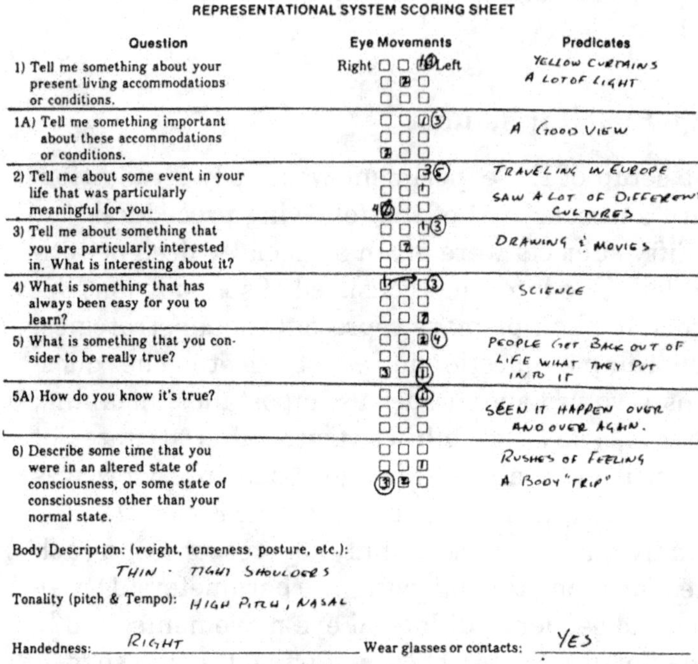

REPRESENTATIONAL SYSTEM SCORING SHEET

Question	Eye Movements	Predicates
1) Tell me something about your present living accommodations or conditions.	Right ☐ ☐ ⊛ Left ☐ ⊠ ☐ ☐ ☐ ☐	YELLOW CURTAINS A LOT OF LIGHT
1A) Tell me something important about these accommodations or conditions.	☐ ☐ ⊠③ ☐ ☐ ☐ ⊠ ☐ ☐	A GOOD VIEW
2) Tell me about some event in your life that was particularly meaningful for you.	☐ ☐ ⊠⑥ ☐ ☐ ☐ ④⊠ ☐ ☐	TRAVELING IN EUROPE SAW A LOT OF DIFFERENT CULTURES
3) Tell me about something that you are particularly interested in. What is interesting about it?	☐ ☐ ⊠③ ☐ ⊠ ☐ ☐ ☐ ☐	DRAWING & MOVIES
4) What is something that has always been easy for you to learn?	⊛ ☐ ⊛ ③ ☐ ☐ ☐ ☐ ☐ ⊠	SCIENCE
5) What is something that you consider to be really true?	☐ ☐ ⊠④ ☐ ☐ ☐ ⊠ ☐ ①	PEOPLE GET BACK OUT OF LIFE WHAT THEY PUT INTO IT
5A) How do you know it's true?	☐ ☐ ☐ ☐ ☐ ④ ☐ ☐ ☐	SEEN IT HAPPEN OVER AND OVER AGAIN.
6) Describe some time that you were in an altered state of consciousness, or some state of consciousness other than your normal state.	☐ ☐ ☐ ☐ ☐ ① ③ ⊠ ☐	RUSHES OF FEELING A BODY "TRIP"

Body Description: (weight, tenseness, posture, etc.): THIN · TIGHT SHOULDERS

Tonality (pitch & Tempo): HIGH PITCH, NASAL

Handedness: RIGHT   Wear glasses or contacts: YES

Figure 13.

noted by the scorers to the side of the appropriate eye-movement matrix. Arrows were drawn, when necessary or practical, from circled eye positions to vocal content.

Overall body type and tonal qualities were judged and described for each individual, as well.

Each interview was videotaped to be saved for future references and scoring verification.

Sometime subsequent to the interview sample EEGs were taken for each subject, as the subject sat quietly with eyes open or closed. Two EEG channels were used, recording activity taken from surface electrode placements on the left occipital (visual) and temporal (auditory) areas of each subject. Two base line readings taken (for both eyes open and closed) at speeds of 2.5 and 5 mm/sec.

These readings were then paired with the scoring made during the interview and correlated.

The results of these evaluations showed a good correlation between base line EEG patterns and primary representational system. Variations occurred when, due to the nature of the questions asked and the individual makeup of the subject, it was not possible to determine a most highly valued representational system.

Some individuals showed a preference for a particular strategy or sequence of representational systems when making decisions as opposed to a single representational system. Correspondingly their base line EEG readings showed more variation in wave shape than subjects who demonstrated a definite preference for one sensory modality.

Other individuals tended to switch the criteria (and representational focus) they were basing their answers on for different questions, indicating a more complex pattern for that person than a single representational system dominance.

Both of these variations have stimulated further refinement and research in the following areas:

1) Determining the most appropriate strategy sequence for specific tasks. For example, good spellers always look up and visualize the letters when they are spelling.

2) Exploring hierarchies of criteria and identifying which representational system is used to determine whether the criteria have been met and how this affects the personality and belief system of the individual.

BIBLIOGRAPHY FOR PART II

Bandler, R., Grinder, J.: *The Structure of Magic I & II.* Science and Behavior Books, 1975, 1976.

Bateson, G.: *Steps to an Ecology of Mind.* Ballantine Books, 1972.

Callaway, E.: Schizophrenia and Interference. *Arch Gen Psychiat* 22: 193–208, 1970.

John, E. R.: *A Model of Consciousness.* 1975

Miller, G.: The Magical Number Seven, Plus or Minus Two. *Psych Review* 83: 81–97, 1957.

Galin, D., Ornstein, R.: Individual Differences in Cognitive Style—Reflective Eye Movements. *Neuropsychologia* Vol. 12: pp. 376–397, 1974.

Dumas, R., Morgan, A.: EEG Asymmetry as a Function of Occupation, Task, and Task Difficulty. *Neuropsychologia* 13: 219–28, 1975.

Kocel, K. et al.: Lateral Eye Movement and Cognitive Mode. *Psychon Sci* 27:223–24, 1972.

Kinsbourne, M.: Eye and Head Turning Indicates Cerebral Lateralization. *Science,* 179: 539–41, 1972.

Grinder, J., Bandler, R., DeLozier, J.: *Patterns of the Hypnotic Technique of Milton H. Erickson M.D.* Vol. I & II. Meta Publications, 1975, 1976.

Pribram, K.: *Languages of the Brain.* Prentice-Hall, Inc., 1971.

⎯⎯⎯⎯⎯: The Holographic Hypothesis of Memory Structure in Brain Function and Perception. 1974.

⎯⎯⎯⎯⎯: Problems Concerning the Structure of Consciousness. *Consciousness and the Brain.* 1976.

Ashby, W. R.: *Design for a Brain: The Origin of Adaptive Behavior.* John Wiley & Sons, 1960.

_____: *Introduction to Cybernetics.* London University Paperbacks, 1964.

Sheldon, W. H.: *The Varieties of Human Physique; An Introduction to Constitutional Psychology.* Hafner, 1940.

_____: *Varieties of Human Temperament; A Psychology of Constitutional Differences.* Harper, 1942.

_____: *Atlas of Men; A Guide for Somatyping the Adult Male at All Ages.* Harper, 1954.

John, E. R.: Switchboard vs. Statistical Theories of Learning and Memory. *Science* 177: 850–64, 1972.

Morrell, F. et al.: Electroencephalography. *Clinical Neurophysiology,* 23, 1967.

Haley, J. (ed.): *Advanced Techniques of Hypnosis and Therapy.* Grune & Stratton, 1967.

_____: *Uncommon Therapy.* W W Norton & Co, Inc, 1973.

PART III: APPLICATIONS OF NEURO-LINGUISTIC PROGRAMMING TO THERAPY

By Robert B. Dilts
1978

TABLE OF CONTENTS
FOR PART III

	Page
Introduction	3
I. Parameters of States of Internal Information Processing	5
1.1 Representational Systems	5
A. Exogenous and Endogenous Activity	6
1.2 Representational System Primacy	6
1.3 Synesthesia Patterns and Lead Systems	8
1.4 Hemispheric Dominance and Lateralization	9
1.5 Limitations of Consciousness	10
1.6 States of Consciousness	12
1.7 Parallel Processing and Parts	13
A. Incongruence	15
B. Unconscious Parts	15
C. Maladaptive Parts	16
II. External Identification of Internal Processes	16
2.1 Accessing Cues	17
A. Lateral Eye Movements	18
B. Breathing Changes	20
C. Muscle Tonus Changes	20
D. Tonality Changes	20
E. Tempo Changes	21
2.2 Physiological Indicators of Primary	

Representational System	21
A. Body Types	21
B. Predicates	22
C. Occupational Interests and Talents	23
2.3 Parts and Splits	24
III. Therapeutic Procedures	25
3.1 Anchoring	25
A. Integrating Parts through Anchoring	27
B. Anchoring Away a Symptom	29
C. Creating Experience through Anchoring	29
3.2 Reframing	31
A. Identifying a Part	34
B. Establishing a Channel of Communication with a Part	38
3.3 The Meta Model	39
A. Information Gathering	40
1. Deletions	40
2. Simple Deletions	41
3. Lack of Referential Index	42
4. Deletion of Comparatives	43
5. Unspecified Verb	43
6. Nominalization	44
B. Limitations of an Individual's Model	45
1. Presuppositions	45
2. Modal Operators of Possibility and Necessity	45
3. Universal Quantifiers	47
4. Complex Equivalence	47
C. Semantic Ill-Formedness	47
1. Cause-Effect	48
2. Mind Reading	48
3. Lost Performative	48
3.4 Altered States of Consciousness: Accessing and Creating Parts	49

Contents

A. Tracking	52
1. Accessing the Appropriate 6-tuples	52
2. Overlapping 6-tuples	53
3. Pacing and Leading	54
a. Pacing	54
b. Leading	54
1) Matching Predicates	55
2) Body Mirroring	56
3) Pacing Breathing and Accessing Cues	57
4) Meta-pacing	58
4. Linguistic Tracking Patterns	58
a. Causal Modeling	58
1) Conjunctions	59
2) Implied Causatives	59
3) Causatives	59
b. Lack of Referential Index	60
c. Nominalizations	61
d. Presuppositions	61
e. Modal Operators, Universal Quantifiers, and Mind Reading	62
f. Lost Performatives	62
g. Tense	62
h. Lesser Included Structures—Analogical Marking and Embedded Commands	63
i. Derived Meanings	65
j. Suggestion and Suggestibility	66
B. Interruption of Conditioned Behavioral Patterns	68
1. Blocking	68
2. Confusion Techniques	70
a. Phonologic Ambiguity	71
b. Syntactic Ambiguity	71
c. Punctuation Ambiguity	71

C. Enhancement and Exaggeration of a Behavioral Pattern	72
3.5 Metaphor	74
A. Isomorphism within Neuropsychological Systems	76
1. Behavioral Generalization	78
2. Classes of Relations	79
IV. Therapeutic Epistemology	81
Appendix A: Instance of Integration of Parts Using Reframing and Metaphor	91
Appendix B: Neurology of Learning	101
Bibliography	108

NEURO-LINGUISTIC PROGRAMMING™

INTRODUCTION

Neuro-Linguistic Programming™ (NLP™) provides a new approach to psychotherapy that is currently spreading in popularity throughout the United States and many areas of Europe. Developed by behavioral modelers John Grinder and Richard Bandler, it is based on their own experience as professional communicators and also draws largely from the therapeutic work of Fredrick Perls, Virginia Satir, and Milton H. Erickson, M.D., founder of the American Society of Clinical Hypnosis. In addition, it is theoretically rooted in principles of neurology, psychophysiology, linguistics, cybernetics, and communication theory.

Neuro-Linguistic Programming does not seem to fall into any one of the existing schools of psychotherapy. It could be considered "behavioristic" in that its major concern is with understanding and utilizing the establishment, interruption, and change of conditioned programs of behavior. It does not, however, operate out of the stimulus-response-reinforcement triad theory of learning and conditioning, because much of its effectiveness is dependent upon single-trial learning without immediate reinforcement. It is based more on the "holonomic" concept of learning and memory (Pribram 1973, 1977). It emphasizes the rules *governing* conditioning and behavior, or conditioning at the level of

Learning II, that is, "learning to learn" (Bateson, 1972).

Neuro-Linguistic Programming is also "psychodynamic" in that it concentrates on the interactions between internal processes or "parts," and on the resolution of conflicts between behavioral programs. Because it is based in cybernetics, NLP focuses on dealing with whole systems in which each part of the system affects and is affected by the other parts of the system. In this sense, a symptom is rarely treated as a problem in itself.

Finally, NLP is "humanistic" in that it presupposes that each individual already possesses the skills and abilities that he/she needs to change his/her own behavior. NLP accepts the fundamental paradox that each human being is nothing more than a "conglomeration of protoplasm on a ball of dust floating through the universe," yet, at the same time, each is in a very real sense the creator of a universe; that a single individual is nothing and yet he/she is all there is.

The purpose of this paper is to give a short, but comprehensive, survey of the principles and uses of Neuro-Linguistic Programming. I will proceed in the following format:

a. Presentation of the core concepts of representational systems and other parameters of human information processing.

b. Presentation of a calculus for the identification of states of internal processing via easily observed external behaviors.

c. Presentation of basic strategies for the facilitation of therapeutic change.

I. Parameters of States of Internal Information Processing

1.1 Representational Systems

As mammals, human beings receive and represent information about their environment through specialized receptors and sense organs located throughout the central nervous system. These perceptual modes can be grouped into five major classes: 1) *vision* (sight); 2) *audition* (hearing); 3) *olfaction* (smell); 4) *gustation* (taste); and 5) *kinesthesis* (body sensations), which are divided into a) *somesthesis* (tactile sensations) and b) *proprioception,* or *visceral* sensations (internal "feelings"). Each class of perception consists of its own system of receptors and transmitters that, depending on its individual structural properties, will "fire" in response to different types and thresholds of mechanical, electrochemical, or electromagnetic irritations in the individual's external and internal environment. Each sensory system is responsible for the transmission and processing of unique sets of distinctions about the individual's immediate universe.

In NLP, each sensory system or *representational system* is considered to do more than simply receive and represent information. Each system also has functional significance in that the distinctions received through each of these systems initiate and modulate (via neural interconnections) the individual's behavior. Action and movement occur as a result of another neural system known as the *motor* system.

Motor movements eventually become established and represented independently from the other systems; but due to the interaction between and feedback from representational systems, the movements are always in some way modulated or programmed by the information being processed through the other systems.

In NLP, behavior can consist of activity in any representational system. Making an internal visual image is as much a behavior as walking. What and how an individual hears would also be considered behavior. External, observable behavior includes all noticeable physiological responses ranging from language, to skin color changes, to breathing rate, to eye movements.

A. Exogenous and Endogenous Activity

Activity in a representational system can be initiated either *externally* by an individual's immediate sensory environment, or *internally* by memory or imagination. (For discussion of neurological differences, see: Dilts, 1977; John, 1975; Pribram, 1973, 1977.) In NLP, it is important to identify and distinguish between behaviors generated by the individual's external setting (or *context*) and the individual's internal response to that context.

1.2 Representational System Primacy

Representational system primacy is the result of the hypertrophy or, in some cases, the atrophy of a particular representational system. Learning, or the establishment of behavioral programs, is determined by the amount of activity within the various representational systems and the interconnections between those systems. In human beings, the senses of taste and smell tend to play a much weaker role in this process, at the organizational level, than do the other senses. An individual planning what to do tomorrow, discussing philosophy, reading a textbook, sewing, or writing will probably not use their sense of taste or smell as an organizational strategy.

During the development of a particular skill or the learning of a particular task, it is often necessary to rely on the information from one representational system to a greater degree than from others. For example, an individual learning to compose music pays attention to a different class of his/her sensory experience than an individual learning chemistry or juggling. In fact, an individual attempting to apply the same perceptual-motor programs developed for learning chemistry (a task that tends to be primarily visual) to the task of learning gymnastics (a task that requires a much greater attention to tactile sensations) would probably have a more difficult time than if he/she had attempted to take up philosophy.

As an individual matures there is also a growth and maturation of brain elements. Many neurons, especially those which integrate the activity of receptors (sensory input) and effectors (motor output), require sensory stimulation for their development. The absence of sensory stimulation during the various critical periods of their maturation produces physiological degeneration of cell activity along a number of parameters including cell size, distance of migration, and electrical response. If, an individual is subjected to internal (genetic) or external (environmental) factors that heavily direct his/her attention to a particular representational system during a critical period of neurological development (the child is handed a violin by his/her mother, rather than a football or paint brush, for instance), one might predict that this will have behavioral implications beyond the span of that critical period. This individual may become conditioned to rely heavily on auditory information for learning or behaving, even in new circumstances or contexts where attentiveness to visual sensations might be more appropriate. This individual may not even know how to make the necessary distinctions (that is, they can *see* but not *observe*).

This process of developing attentiveness to distinctions within a particular class of sensory experience, beyond task-specific purposes, is an example of the process of learning to learn or Learning II. It is obvious that which portions of sensory experience an individual pays attention to as he/she attempts to learn something will greatly enhance or hinder the pace and effectiveness of learning.

1.3 Synesthesia Patterns and Lead Systems

In NLP, *synesthesia* describes the cross-modal or bi-modal representation of a particular experience. An example of *cross-modal* synesthesia would be hearing a harsh tone of voice and feeling uncomfortable in the stomach area as a result. An example of *bi-modal* synesthesia would be experiencing colors as you listen to music, seeing a sound. It is clear that these experiences are the result of crossover connections between representational systems in the brain in such a way that activity in one representational system can initiate activity in another. These patterns constitute a large portion of human meaning-making processes.

It is obvious that synesthesia patterns may also become overconditioned. An individual may get strong internal feelings from everything he/she sees. Often an individual will develop the following pattern: Something that they see externally, or visualize internally, will initiate an internal kinesthetic response; that kinesthetic activity, in turn, initiates a motor response or action. For instance, a person sees blood, begins to feel nauseated, and turns away. Or a person sees a sunset, feels elated and sighs. In the NLP model, the system through which the information is received is called the *lead system;* the system that initiates behavior is called the *primary representational system.* In general, information is received through an individual's lead system and processed in the primary representational system. Depending on the context, the two may, or may not, be the same sensory modality.

1.4 Hemispheric Dominance and Lateralization

In human beings, each hemisphere of the brain controls almost exclusively the motor activity on the opposite side of the body: The left hemisphere controls the right side of the body, and the right hemisphere controls the left side of the body. Because it is often more efficient to specialize the coordination of many activities to one side of the body (handedness, for example), one hemisphere may become "dominant" in that its activity controls the dominant side (the side most often used). This is another example of the hypertrophy of activity.

Although there is considerable overlap of the sensations from the eyes and ears between the two hemispheres through the corpus callosum, the representational system governing the activity of the dominant-hemisphere functions have a tendency to lateralize to a degree as well. Most language functions are typically localized in the dominant hemisphere.

Typically, a person's primary representational system will lateralize to the dominant side while another representational system may mediate the motor activity of the other hemisphere. For example, we have probably all heard someone say, "On the one hand this really *looks* like it would be fun, but on the other hand, it just might not *feel* right," (or ". . . something *tells* me that I really shouldn't do it").

There is also evidence to support the idea that the hemispheres actually carry out different functions: The dominant hemisphere (the left hemisphere in most right-handed people) carries out sequential, cause-effect reasoning, such as language and logic; the non-dominant carries out spatial, integrative, or "artistic" types of processing. Many of these properties can also be tied to the functions of different representational systems. (Incidentally, it has been shown by recording electrical brain activity that an individual does not "live" in one hemisphere or the other; rather, activity is lateralized in response to specific tasks. So, electrical brain activity is generally asymmetrical less than 20% of the time, even during problem-solving tasks (Dumas and Morgan, 1974).

1.5 Limitations of Consciousness

Consciousness is a concept that often has a multiplicity of meanings. In Neuro-Linguistic Programming, the phenomenon of consciousness is specified as the act of being aware of an internal or external sensory stimulus. Fundamental to the NLP definition is that consciousness is an emergent property of structural psychodynamic activity rather than an entity in itself. To say that the property of consciousness controls or affects behavior is like saying that the properties of wetness or iciness controls or affects the structural combinations of H_2O molecules. Rather, in NLP, consciousness is considered to be the result of varying signal-to-noise ratios of activity within the various representational systems. The signal within a particular representational system is determined by the interaction between the intensity of the external stimulus and the internal brain state, at a particular point in time.

Because of its nature, it is obvious that consciousness is a limited phenomenon: an individual cannot pay attention to all his/her incoming sensory experience at once, although he/she may be able to respond to it automatically. Psychological tests (Miller 1957) place the limits of conscious attention at 7 ± 2 "chunks" during the performance of a task. (*Chunk* refers to levels of a particular cognitive gestalt of information—a chair may constitute one level of chunking and a chair leg another; a movement one level of chunking and an action another. In general, the smaller the chunk, the greater the detail.)

It seems to follow that an individual's primary representational system will have the highest signal, or be "most conscious," for that person—unless the programs mediated by that system have become habituated (when the individual stops "learning" in that system and the responses become automatic). In such cases, an individual's lead system may be his/her most conscious system. For instance, I often hear people say, "I *feel* that I'm a *visual* person," or "I *see* myself as *auditory*," or "Something *tells* me I'm primarily a *feeling* person."

In NLP, notions of self and volition are also considered emergent properties. Choices and decisions are constantly being made about all aspects of behavior. It is only when one is aware of the process that he/she claims to be in "control" of them.

Unconscious behavior or an individual's "unconscious mind," is made up of stimuli within a particular representational system that initiates a behavioral response, but of which the individual is typically not conscious (i.e., has a low signal).

1.6 States of Consciousness

A person's state of consciousness is made up of the interplay between contextual conditions (type and intensity of stimuli in the individual's environment), the portions of that context that initiate behavior (the individual's lead and primary representational system at that increment in time), and, of course, those parts of the interaction that the individual is most aware of (the amount of signal produced in a certain representational system). An *altered state* occurs when one of the parameters is changed. For example, individuals who feel they have entered what is called a psychic state will describe their normal state of consciousness as *leading* visually and *representing* kinesthetically (typically, getting feelings from what they see). In the psychic state, however, they *lead* with internal visceral feelings, and the *representation* (the sensations of primary behavioral import) becomes the internal visual image that they make from that feeling.

In this definition note that there is also a formal difference between a *state of consciousness,* as described above, and a *state of awareness*. In a state of awareness, context, lead, and representational systems remain constant, but the amount of signal (or conscious awareness) is displaced to a different portion of the sensory spectrum.

1.7 Parallel Processing and Parts

The concept of *parts* is fundamental to the psychotherapeutic techniques and goals of Neuro-Linguistic Programming. A part is essentially a persona, identifiable by various lead and representational system combinations, that evolves from the different behavioral responses and programs developed by an individual in different contexts and states of consciousness. The strength of a part is dependent on the intensity, duration, and consolidation of the environmental stimulus, the resulting program, and the state of consciousness.

The neural propensity for the establishment of hemispheric dominance and the dissociation of functional domains of different brain areas and systems is also of great importance to the establishment of parts. These mechanisms predict the simultaneous processing of the same sensory stimulus independently, to some degree, by different parts—either in terms of hemispheres or separate representational systems. Ashby (1960) and Pribram (1969) give accounts of how, in organisms and artificial intelligence, multiply interconnected information-processing mechanisms can become disjoined and operate as separate channels when parts of the system come under control of separate environmental inputs.

One outcome of the brain's ability for parallel processing is the possible competition between multiple behavioral responses to the same external stimulus. For instance, when there is no transference between representational systems, a particular experience may literally *look* different than it would *feel*. Callaway (1969) interprets some schizophrenic behavior as the result of interference within an individual, caused by mismatched responses to external stimuli, and the efforts of the individual to cope with that interference.

Because of the cybernetic nature of the neural interface at which these conflicts take place, the results are not limited to cognitive activity within particular representational systems, but extend to motor responses, and even to the regulation of autonomic activity. Mind (neural processes) and body are part of the same biological system. The activity in each affects and is affected by the activity of the other (as anyone who has ever been in serious conflict within him/herself will understand). Conflict between parts can directly or indirectly affect information processing in other neural systems (synesthetically), autonomic control of glandular secretions, body chemistry, heart and blood pressure rates, respiration, basal metabolism, and even the immune system. The conflicts may not directly *make* these changes, but any neural activity can't *not* have some effect on the rest of the system.

For NLP, so-called psychosomatic illnesses are not "all in the mind," but are the result of a very real interaction between biological mechanisms. It is a goal of NLP to identify and resolve conflicts of this nature.

A. Incongruence

The mismatch of behavioral responses caused by conflicting parts is called *incongruence*. Parts may express themselves either *simultaneously* or *sequentially*. Sequential incongruence would be characterized by the case of an individual who says, "I'd really like to go, *but* I really feel the need to stay home." An individual who is simultaneously incongruent would say, "Yes, I'd really like to go," but would be shaking his/her head "no" at the same time.

Some techniques of NLP are designed to sort and identify conflicting parts sequentially: first, so that the responses will be instantaneously congruent; and then to reintegrate the parts simultaneously in such a way that the conflicting responses may be resolved.

B. Unconscious Parts

Due to the inherent limitations of consciousness, a particular part may contain more signal at a particular point in time than another, although both are operative simultaneously. People are often unaware of their own behavior. An unconscious part is most likely to be the result of the person's least conscious representational system. Often, the most habitual behavior is initiated by the least conscious system.

Another result of the limitations of consciousness in parts is when the part's lead system (the system that initiates the response) is outside of the individual's awareness. In these cases, the individual's conscious mind finds itself feeling or responding in a certain way without knowing how or why. Intuition is an example of a representation or response that has been dissociated from the initiating information.

It is not necessary, in NLP, to make a person conscious of all of his/her behavior. Consciousness is certainly not tantamount to change. In fact, a person's unconscious behavior can often be utilized as an extremely powerful tool for change.

C. Maladaptive Parts

As I pointed out earlier, an adaptive response to a particular stimulus in one context may become maladaptive in a different context. For instance, when you are a child, responding to a certain tone in your mother's voice by withdrawing may be very adaptive given the relationship between a parent and child. Responding with the same reaction at a later point in your life, or to that same tone from your wife or husband, however, may lose adaptive significance given the new context. The constant reordering and restructuring of behavioral responses with regard to changes in context is another fundamental purpose of NLP.

II. External Identification of Internal Processes

The purpose of this section is to present the reader with a means of identifying the interplay between lead and representation within an individual through the identification of external behavioral transforms of that activity. Although I will not go into any lengthy explanations or justifications for the parameters I have chosen, I will cite supporting clinical studies when appropriate.

The following are two basic principles underlying this sort of classification of behaviors:

a. Any occurrence in one part of a cybernetic system (such as a human being) will necessarily affect all the other parts of that system in some way. When the rules of interaction between the parts of the system are understood the effects of the different parts of the system on one another can be patterned, predicted, and changed.

b. In animals, all behavior (from propulsion to pigment changes to respiration changes to acts of communication) is a transform of internal neural processes, and therefore carries information about those processes. It follows, then, that all behavior is in some way communication (that is, an organism can't *not* communicate or respond).

2.1 Accessing Cues—Identifying Lead Systems

When an individual is asked to recall some bit of information that is not available in his/her immediate sensory environment, he/she must go through the process of accessing that information, either through memory or construction. To assist in this process, individuals establish behavioral cues ranging from a finger snap to positions of lateral eye movement to mutterings and utterances to breathing rates and positions. (We have probably all observed these behaviors in an individual searching for the answer to a question.) Similarly, as an individual is organizing his/her experience for him/herself, that individual will also use these cues. One can observe and identify patterns for these cues in response to the accessing of information of different representational modalities and hemispheric processing.

A. Lateral Eye Movements (see Figure 1)

The research of Kinsbourne (1972, 1973), of Kocel et al. (1972), and of Galin and Ornstein (1974) supports the hypothesis that the direction of lateral eye movement is associated with the access or activation of the contralateral cerebral hemisphere in response to varying kinds of cognitive tasks. NLP research has expanded the scope of this type of research to include not only left-right lateral eye movement, but also vertical eye position (up-lateral-down) in response to sensory-specific questions—Dilts (1977), Owens (1977), Thomason, et al. (1980). Putting this research together with clinical observation, the following patterns have been identified:

1. *Up and left*—accesses non-dominant hemisphere visualization (for right-handed people): remembered or eidetic imagery.

2. *Up and right*—accesses dominant-hemisphere visualization: constructed imagery and visual manipulation.

3. *Level and to the left*—accesses remembered sounds and tape loops (for example, nursery rhymes): non-dominant hemisphere tonal discriminations.

4. *Level and to the right*—auditory construction and manipulation (dominant hemisphere): thinking of things to say, syntax, etc.

5. *Down and left*—internal dialogue.

Neuro-Linguistic Programming

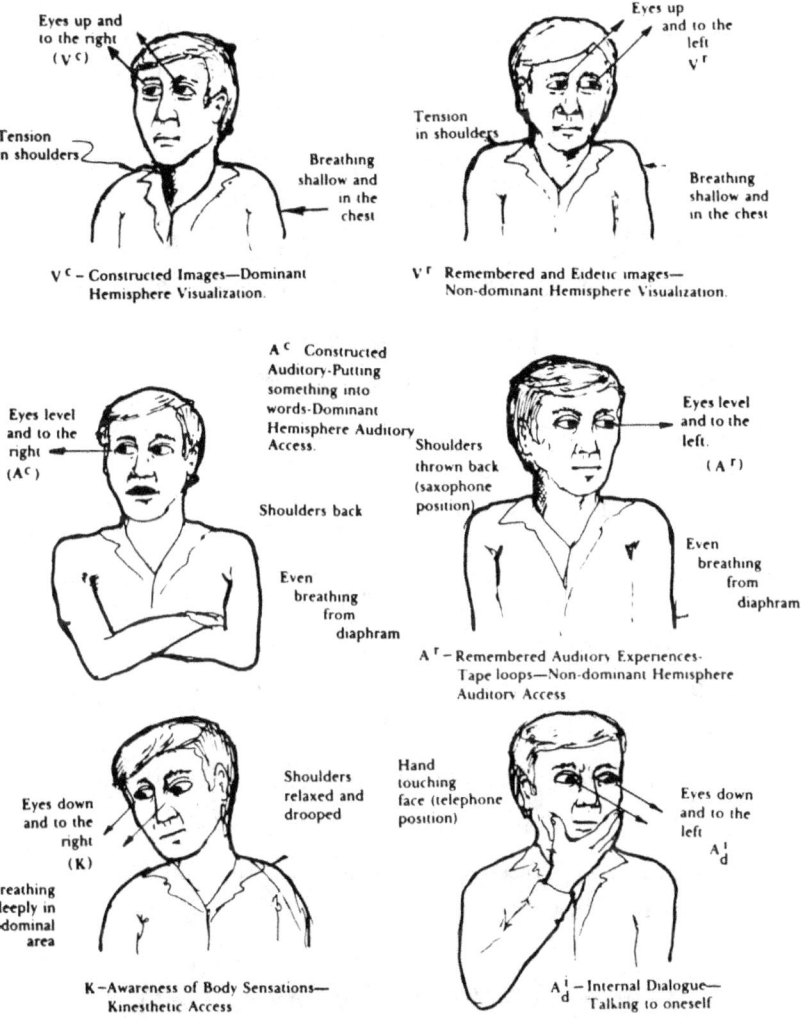

Figure 1. Accessing cues for typically wired right-handed person.

6. *Down and right*—accesses both tactile and visceral feelings.

7. *Straight ahead, but unfocused and/or dilated*—

quick access of almost any class of sensory information: generally indicates the representational system the person can access most easily.

B. Breathing Changes

1. Breathing *high and shallow in the chest* or the cessation of breathing indicates visual accessing.

2. *Deep breathing low in the stomach area* indicates tactile or visceral kinesthetic accessing.

3. *Even breathing in the diaphragm* or using the whole chest, with a *typically prolonged exhale,* indicates auditory accessing.

C. Muscle Tonus Changes

1. Increase of *muscle tension,* particularly in the *shoulders and abdomen,* is indicative of visualization.

2. *Increase of movement* indicates tactile kinesthetic accessing.

3. *Muscle relaxation* indicates internal-visceral kinesthetic accessing.

4. *Even muscle tension and minor rhythmic movement* generally indicates auditory accessing.

D. Tonality Changes

1. *High-pitched, nasal and/or strained tonality* is indicative of the access of visual information.

2. *Low, deep tonality* indicates kinesthetic access. The voice is often more *breathy.*

3. *Clear, precise, resonant* tonality with a lot of *variation* indicates auditory access.

E. Tempo Changes

1. *Quick bursts of words* and a generally *fast tempo* indicates access of visual information.

2. *Slow tempo with long pauses* indicates kinesthetic access.

3. *Even rhythmic tempo* is characteristic of auditory access.

2.2 Physiological Indicators of Primary Representational System

Any of the accessing cues listed previously, if exaggerated or used extensively over time, will come to indicate one of the individual's primary representational systems. An individual with a typically high-pitched voice, who breathes high and shallow in the chest, will tend to be visually oriented. An individual with a typically low voice and slow tempo will tend to be more kinesthetic. Further, these accessing cues carried out over long periods of time will begin to shape an individual's body, contributing to the possible hypertrophy and/or atrophy of various muscle groupings, and affect his/her metabolism. If an individual has a constant tenseness in his/her shoulders for a long period of time, it will show up in that person's posture.

A. Body Types

An individual's body type is the result of the interaction between genetic factors and the kinds of processes mentioned above that tend to shape the anatomy. The body typology proposed by NLP tends to correspond, basically, with the research done by Sheldon. The endomorph (a soft, fleshy body) characterizes an individual whose primary representational system is visceral kinesthetic; the mesomorph (athletic and muscular) indicates an individual more tactually oriented; the ectomorph (thin and tense) corresponds to a more visually oriented individual. Sheldon skips the auditory individual (probably his own least conscious representational system). This representational system tends to show up in posture, and voice quality in most instances, anyway.

1. A *thin tense body* is characteristic of a visually oriented person.

2. A *full soft body* is characteristic of a visceral-kinesthetic person.

3. An *active muscular body* characterizes an individual that is more tactually sensitive.

4. The body of an auditorily oriented person tends to be in between that of a visually oriented person and that of a kinesthetically oriented person. (The body will tend to be softer if the individual is more attentive to internal auditory information; more tense if the focus is external.)

B. Predicates

Bandler and Grinder (1975, 1976) have pointed out that the predicates an individual uses to describe his/her experience will be an accurate transform of the way that experience is represented to the individual. An individual that tends to use a preponderance of visual predicates will probably be more aware of the visual aspects of his/her experience; one who uses auditory terms, more aware of sounds; and so forth.

1. *Visual:* I *see* what you are saying; That *looks* good; That idea isn't *clear;* I'm *hazy* about that; I went *blank;* Let's cast some *light* on the subject; Get a new *perspective;* I *view* it in this way; *Looking back* on it now; An *(enlightening, insightful, colorful)* example.

2. *Kinesthetic:* If it *feels* right, do it; Get a *handle* on it; *Grasp* the concepts; Get *in touch* with yourself; A *solid* understanding; I'm up *against a wall;* Change your *standpoint;* You're so *insensitive;* I *have a feeling* you're right.

3. *Auditory:* I *hear* you; That *rings a bell;* It *sounds* good to me; Everything just suddenly *clicked;* *Listen* to yourself; That idea's been *rattling* around in my head; Something *tells* me to be careful; I can really *tune in* to what you're saying.

C. Occupation, Interests, and Talents

Individuals who have developed a particular skill or attentiveness to a particular representational system will tend to sort themselves occupationally in terms of what they enjoy and what they are good at. The skills that an individual cultivates, then, will also be a transform of their internal skills. Some possible examples include:

1. Visual—Engineering and drafting, the visual arts (painting, drawing, etc.), sciences (physics, mathematics, chemistry, etc.).

2. Kinesthetic—Athletics, tactile arts (sculpture, ceramics, etc.), manual tasks (construction, housekeeping, cooking, etc.).

3. Auditory—Music, literary arts (writing, poetry, etc.), linguistics and languages.

Occupations and arts are often varied in terms of what sensory system they appeal to most. Different schools of thought within a particular occupation may be indicative of different representational approaches. An art therapist or a Jungian therapist may tend to be more visually oriented than a therapist who does Rolfing or Riechian body work. A psychoanalyst or Rogerian therapist may tend to be more auditorily oriented.

2.3 Parts and Splits

Every individual, of course, has a number of combinations of parts, covering a number of representational systems. Often the physiological indicators mentioned here will change, as a particular part becomes dominant (depending on the time and intensity of dominance). Someone who becomes more in touch with his/her body may gain weight, for example. An anorexic, on the other hand, may care only for how she looks.

Individuals with strong parts that are coexistent may establish combinations of representational indicators. A person who has strong visual and kinesthetic (visceral) parts and who typically leads visually and represents kinesthetically, will tend to have a high-pitched voice, to speak quickly, to move his/her eyes upward consistently, and to have tense, thin shoulders. They will also tend to use kinesthetic predicates to describe their experience, and be heavy in the stomach and hips (their body takes on a pear shape).

III. Therapeutic Procedures

3.1 Anchoring

In its most general sense, an anchor is any stimulus that elicits a consistent response. The major difference between the concept of anchoring and stimulus-response theories of behavior is that the establishment of an anchor does not require reinforcement. In this sense, anchoring is more of an extension of the holographic, or holonomic, theory of learning and memory (Pribram, 1969, 1973, 1977).

The basic theory of anchoring maintains that any experience is represented as an individual gestalt of sensory information. When that information is recorded against a steady, coherent source (a brain state, in this case), whenever one portion of the original cognitive gestalt is reintroduced, all of the other components of that experience will be reproduced to some degree. This process happens as a result of the synaptic and electric interference patterns created during neural processing. Its form is similar, in many ways, to an optical hologram. Any part of a particular experience may be used to access another part.

One interesting aspect of this phenomenon is its dependence on the brain state. The anchors established during one state of consciousness may lose effectiveness in another state. In fact, the stimulus may anchor something completely different in the other state. Further, in a state of consciousness that is entered rarely by an individual, there will be less interference from previous responses to typical contextual anchors. The individual may be able to learn more rapidly, therefore, in the absence of interference and literally experience stimuli as having new meaning.

A particular signal or stimulus is only meaningful to an individual in as far as it elicits a response from that individual. The word "dog," for example, has meaning for the reader only if that arrangement of letters accesses an internal representation(s) from the reader's past sensory experience. The representation may change if I change the context by adding the anchor, "wet" before "dog."

Anchors may be established through any sensory modality: a touch constitutes as effective an anchor as a written or spoken word. It is obvious that the reinforcement of an anchor *will* generally tend to increase its effectiveness, although this is not always the case. Often it is the initial experience that establishes the anchor most firmly.

A particular anchor can be represented as a *6-tuple*. An anchor consisting of a spoken word would look like this:

$$A_d[\ A_t^{e,i}, K_t^{e,i}, V^{e,i}, K_v^{e,i}, O^{e,i}, M\ ]$$

In this example, the symbol A_d stands for the spoken word. The subscript, d, identifies the stimulus as being *digital* as opposed to tonal, A_t. The superscript, e,i, for each of the representational classes indicates that the experience within that modality may be from either *external* or *internal* sources. The symbols within the brackets should be understood as follows: A_t—*auditory tonal;* K_t—*kinesthetic tactile;* V—*visual;* K_v—*kinesthetic visceral;* O—*olfactory;* (Gustation may be added also); M—

motor activity. The representation means that for the stimulus A_d some set of the representational systems listed will be accessed. These will vary in intensity, duration, frequency etc., depending on the individual's state of consciousness. The A_d component may, of course, be replaced by any other sensory stimulus.

Accessing cues are examples of self-established anchors. Lead systems are also the result of this process.

The phenomenon of anchoring is one of the Neuro-Linguistic Programmer's most powerful tools.

A. Integration of Parts through Anchoring

Anchoring conflicting parts and integrating them is one of the simplest and most powerful techniques in NLP. As the client expresses his/her conflicting parts, the programmer separates them sequentially by asking the client about various aspects of the conflict. As the programmer begins to be able to identify the part in the client's response (by watching accessing cues and listening to predicates), he/she may reach out and anchor that part by a slight touch on the knee. When the programmer identifies the other part that is contributing to the conflict, he/she may reach out and anchor the other part on the opposite knee, or elsewhere on the client's body. It is not necessary that the client be unaware of the touches, as long as they neither interfere with nor interrupt the client's responses.

The programmer may continue to reinforce the anchors if he/she wishes. At some point, however, all the programmer need do is to reach out with both hands and touch both of the anchors simultaneously (this may be done at the same time the programmer offers a suggestion or resolution). In my experience, the response to this simultaneous anchoring is always powerful. Depending on the parts, and the gravity of the conflict, the result may range from tears to convulsions to laughter to momentary confusion.

It is hoped that the simultaneous anchoring will force the two patterns of behavior into the same time and space, neurologically, so that a third pattern of behavior can be formed. Using our notational system we can show the anchoring process in the following way:

$$\female_1 \to [A_1, V_1, K_{t1}, K_{v1}, O_1, M_1]$$
(touch on left knee) → *(part/experience #1)*

$$\female_2 \to [A_2, V_2, K_{t2}, K_{v2}, O_2, M_2]$$
(touch on right knee) → *(part/experience #2)*

We would show the integration process as:

$$\female_1 + \female_2 \to [A_{(1*2)}, V_{(1*2)}, K_{t(1*2)}, K_{v(1*2)}, O_{(1*2)}, M_{(1*2)}]$$

This shows that at the simultaneous firing of the two anchors the two 6-tuples combine to make a third. The star function "*" represents a function of combination of the sensory parameters of the two states. The manner in which this combination takes place will, of course, not be simply additive but will depend on the strength and content of the representations and the neurological attributes of the individual.

B. Anchoring Away a Symptom

In this process an anchor is used to establish a synthesia pattern with a symptom in such a way that when the anchor is substituted for or changed, the symptom must necessarily disappear. I have often used this method for pain control such as treatment of tension and migraine headaches.

In the case of a headache, the client is instructed to make a picture of the pain. If, and when, the client is able to do this (sometimes sounds, or other feelings, work better for an individual), he/she is then instructed to make the image more and more vivid until the headache (or other pain) begins to hurt slightly worse. In this way the image becomes an anchor for the pain. The client may be asked to repeat this a few times to reinforce the anchor. The client then is instructed simply to change the picture to something pleasant. In most cases this will be enough to either reduce or dissipate the pain, and supply the client with a technique for further pain control on his/her own.

We can represent this process the following way:

$$V^i \rightarrow K^{ie}$$

(vividness of image) → (intensity of pain)

As we vary or change the qualities of the image, we vary or change the feeling.

C. Creating Experience through Anchoring

In the same way that parts may be integrated by anchoring them together, various aspects of separate experiences may be anchored together literally to create a new experience for an individual. For instance, the programmer may be working with a client who consistently feels uncomfortable in group situations. The programmer may begin by asking the client to think of a time when the client *did* feel comfortable (regardless of context). The client then is instructed to concentrate on just the feelings of comfort (the feelings are dissociated from their context). The programmer then anchors these feelings.

$$♀_1 \rightarrow [K_{t1}, K_{v1}]$$

Next, the client is asked to recall a time when he/she was uncomfortable in front of a group. The client is directed then to pay attention to just the auditory and visual aspects of the situation. (This can be done by asking the client specific questions about the way things looked and what people said.) For this experience, the context is anchored and dissociated from the feelings.

$$♀_2 \rightarrow [A_2^e, V_2^e]$$

Finally, the group context and the comfortable feelings are anchored together. Then, as the programmer continues to hold the two anchors in unison, the client is asked to recreate the experience as it would have gone, had the client felt comfortable.

$$♀_1 + ♀_2 \rightarrow [A_2^e, K_{t1}, V_2^e, K_{v1}]$$

3.2 Reframing

Reframing is more complex than anchoring and, in fact, involves a good deal of anchoring on the part of the programmer.

The underlying principle of reframing is that *all behavior is or was adaptive, given the context in which it was learned.* In this way, negative or maladaptive behaviors can be reframed as having a *positive intent.*

The process of reframing proceeds in a fairly step-by-step manner:

1. The client and programmer work together to *identify and/or anchor the behavior that has become maladaptive.* This behavior can range anywhere from a backache to a tumor to a feeling of unease around a certain individual to a compulsive habit.

2. The programmer then works to identify and to *separate the context* (the external factors) *from the response to that context* (the internal response) and to anchor the internal response. Generally, this is done auditorily by calling the response a "part."

For example, "I understand that in situation X, part A responds by doing behavior A."

3. *A channel of communication,* or response, *is set up with the anchored part* (this is done best in conjunction with step 4). The response may take the form of an internal voice, a slight twitch, a feeling, an intensification or waning of the problematic behavior, and so forth. There are as many possible channels of communication as there are parts.

If the programmer meets with resistance or a lack of response, he/she can reframe an absence of response as a communication.

For example, "If that part's answer is 'yes,' I would like it to respond with no changes in your present state; if the answer is 'no,' I would like it to respond by increasing behavior M."

4. *Separate the intent of the part from the maladaptive response.*

For example, "Go inside yourself, contact part A, and ask it what it is trying to do for you." Sometimes you may initially get a negative response like, "It's trying to kill me;" in such instances continue searching for a positive intent by asking, "what is it trying to do for you by killing you?"

5. *Make sure all of the other parts of the individual involved understand, and accept, the positive intent of the problematic behavior.* (This can also be done by having the client "ask inside.")

6. *Identify the desired change in behavior and at least three ways to implement that change.* This need not take place at the conscious level. An individual may need to make a number of profound unconscious changes before they can even begin to implement the desired behavior. In many cases other parts, or behaviors, will also need to be reframed during this process. (If this occurs, begin again with the new part at step 3.)

The development of new strategies or behaviors to be substituted for the maladaptive ones can often be assisted by anchoring a "creative part." (This can be done simply by asking the individual to think of various times he/she has been creative, and anchoring them.)

The programmer can also have the individual identify someone that he/she knows who already behaves the way that the client would like to, and have the client observe and model that person's behavior.

7. *Have the formerly problematic part accept and support* (or at least be willing to try) *the new strategy or behavior.*

If the part will not accept the new strategy, find out if there is more important information that needs to be communicated before the part's intention is completely understood; or, find out what the part wants and needs before it will be able to accept the new terms.

8. *Ecological check*: Make sure it is OKAY with all of the person's other parts to change the former behavior to the new one. If any other part objects, reframe the objection by repeating the process.

9. *Future pace*: The client is assisted in constructing, or reconstructing, a new experience based on their new program of behavior. This may be done by using anchoring technique C, described earlier, or by setting up the situation in question in the client's immediate environment, if possible.

Reframing, and the brief techniques of anchoring mentioned earlier, are methods of reprogramming an individual by identifying and reordering, or re-establishing, the anchors that make up and initiate programs of behavior. In reframing, the programmer uses anchoring in two different ways to carry out the two most essential parts of the process:

a) The maladaptive part, or response, is reframed by anchoring the part independent of context. The part then is held steady while various contexts are introduced until the context is determined for which the response is appropriate.

b) New strategies of behavior are determined by anchoring the context in which the former response was problematic, holding that context steady, while accessing and introducing other responses, parts, or combinations thereof available to the individual until the appropriate combination is determined. The context and newly devised response are then anchored together, so that a new program is established.

The two most common obstacles to the programmer carrying out a successful reframing are: a) identifying and gathering information about a particular part; and b) establishing a means of communication with that part.

A. Identifying a Part

In the next section of this paper, 3.3, I will present an explicit set of techniques for gathering information by attending to classes of words (verbal anchors) used in natural language: the Meta Model. This model used in conjunction with nonverbal indicators, such as accessing cues and representational system characteristics, constitutes one of the most powerful sets of information gathering tools available to any professional communicator.

The goal of this process of information gathering is to decode the observable external transforms of internal neurological processes in order to get some understanding of how they interact. In this sense, the programmer can utilize any observable behavior that is coincident with those processes. A particular body posture may anchor a particular state of consciousness or indicate a particular part, for instance. If the programmer suspects some pattern such as this, he/she may verify it by simply having the client reassume that posture, and report on any changes in his/her internal experience. (See Appendix A for a specific example of this.)

The process of discovering the regularities between an individual's overt behavior and his/her internal processes is called *patterning*. Gregory Bateson (1975) formalized some of the properties of this process:

> If from some perception X, it is possible to make better than random guesses about some Y, there is *redundancy* between X and Y, or "X is a coded message about Y," or "Y is a transform of X," or 'X is a transform of Y.' ... When an observer perceives only certain parts of a sequence or configuration of phenomena, he is in many cases able to guess, with better than random success, at the parts which he cannot immediately perceive (guessing that a tree will have roots, for example).

The ultimate success of a programmer will depend on his/her ability to observe and identify the multitude of patterns and transforms his/her client constantly offers. The programmer's abilities to be completely sensorily aware of his/her client are of extreme importance to his/her therapeutic success. The more sensory experience a programmer has of another individual, the more information he/she has to help him/her assist that person in changing.

One method a programmer may use to get information systematically about a client's abilities to organize his/her experience is to ask the client to perform a number of simple sensory-specific tasks (generally in the form of responses to questions). The tasks may appeal to each sensory modality in turn, and also to differing hemispheric processing types (memory versus construction being the most basic distinction). In this way the programmer gathers information concerning: a) which representational systems and processes the client finds most or least difficult, b) which overt behaviors the client will exhibit as he/she momentarily enters the state of consciousness necessary to access the required information, and c) typical combinations of lead systems and accessing cues the client may use to access various parts. Examples of the latter include: One individual may consistently lead visually to get information about his/her feelings (to *see* what he/she look like as his/her are *feeling* a certain way) as well as to get information about past visual and auditory experiences; another individual may lead kinesthetically (with some body movement) to access visual information; another, lead auditorally to access visual information; yet another, lead visually and kinesthetically to construct auditory experience, and so on. Some individuals may be able to access remembered sounds immediately, but may need to lead visually in order to construct auditory experience. Some individuals will have extremely consistent leads, others will vary immensely in response to specific tasks.

Some sample questions of varying complexity include:

1. *Visual Remembered:* What is the color of your car? Who were the first five people you saw today? What kind of pattern is on your bedspread? What is the color of your mother's eyes?

2. *Visual Construction:* Imagine an outline of yourself as you might look from six feet above us and see that image turning into a city skyline. Can you put

together the top half of a toy dog with the bottom half of a green hippopotamus?

3. *Auditory Remembered:* Can you think of one of your favorite songs? What were the last words spoken to you before you arrived here today? Think of the sound of clapping. How does your car's engine sound?

4. *Auditory Constructed:* Imagine the sound of a train's whistle as it turns into the sound of pages turning. Can you hear the sound of a saxophone and the sound of your mother's voice at the same time?

5. *Kinesthetic (tactile) Remembered:* When was the last time you felt very wet? Imagine the feeling of snow in your hands. What does a pine cone feel like? When was the last time you felt a hot cooking utensil?

6. *Tactile Construction:* Imagine the feeling of stickiness turning into the feeling of sand shifting between your fingers. Imagine your hand touching dog's fur and soft butter at the same time.

7. *Kinesthetic (visceral) Remembered:* Can you think of a time you felt satisfied about something you completed? Think of what it feels like to be exhausted. When was the last time you felt impatient? What is it like to be really comfortable about what someone is saying?

8. *Visceral Construction:* Imagine frustration turning into the feelings of being truly motivated to learn something. Imagine the feelings of being bored turning into feeling good about feeling bored.

These questions, of course, appeal to internally generated experience. To get an understanding of the client's *external* sensory skills and abilities, the programmer may ask: What are you aware of right now? Can you hear the traffic outside? How many different colors can you identify within the object you're staring at right now? How do you feel sitting in that chair right now? How do you feel about those feelings? and so on. As the individual answers these questions the programmer also gathers information about how the individual behaves as he/she pays attention to different classes of external stimuli.

B. Establishing a Channel of Communication with a Part

Again, in this procedure, the programmer's sensory awareness of the client cannot be overemphasized. The programmer's ability to perceive minute changes in the client's behavior, as he/she addresses a communication to one of the client's "parts," will greatly enhance his/her ability to establish a channel of communication with that part. The response from a part may be in the form of a muscle twitch, a skin color change, an itch, a body posture, a change in tonality, a change in breathing rate, and so on. Many times these behaviors will occur outside of the client's conscious awareness.

Sometimes the programmer may bring these responses to the client's attention, although depending on the content of the communication, the programmer often may want to respect the unconscious form of the channel of communication and merely acknowledge the completion of a communication to the client's other, more conscious, parts.

The programmer may verify the establishment of

a particular channel by simply noting patterns of consistency between the response received through that channel and communications directed toward that part. (Bringing the response of a particular part to an individual's conscious attention will involve, in actuality, the introduction of the stimulus being processed by that part to a different representational system.)

Establishing communication with a certain part is also enhanced by *pacing* that part: that is, by identifying and respecting the representational integrity of that part. If the programmer observes that the part in question consistently processes and represents information auditorily that is initiated visually then the programmer may gear his/her communications to account for such processing (by not expecting a heavily kinesthetic part to be able to *see* the results that it is causing, for example). By keeping his/her verbal and nonverbal anchors consistent with the part's representational abilities, the programmer can also greatly reduce any apparent resistance by the part.

3.3 The Meta Model

The Meta Model is based on the premise that words (*surface structure*) are meaningful only to the extent they anchor some internal sensory representation or experience (*deep structure*) in an individual. Important information can be lost or distorted, during the codification of experience into verbal anchors (as an individual generates language) and the process of decoding (as the auditory or written stimulus is transformed into internal representations by a second individual). Deletions and distortions of experience may also occur as an individual establishes and orders anchors, using language, for their own experiences. (Cultural and local standards of vocabulary and syntax will contribute to the formation and limitation of the development of verbal anchors and the experience they represent.) It is obvious that words can anchor programs of behavior (6-tuples with a motor, "M", component) as well as simple representations (for example, "Watch out!"). The way an individual orders and sequences these programs through language can be an important determinant in his/her behavior.

In *The Structure of Magic I* (1975), John Grinder and Richard Bandler identify a number of classes of verbal anchors and syntactic environments (the context in which the words appear) that could become problematic during the course of communication. They also provide a series of therapeutically oriented responses that the programmer may use when he/she identifies one of the patterns in question. The effect of the Meta Model is twofold: (1) it will help insure more complete communication between the programmer and the client; and (2) by directly confronting and dealing with the anchors that an individual has developed for his/her own experience and behavior, the client may be able to make rapid and profound changes in behavior by simply "talking" about it.

A. Information Gathering

1. *Deletions:* Deletions occur whenever an anchor for some object, person, or event has been left out of the surface structure that an individual is using to describe his/her experience. A deletion is well-

formed when there is sufficient redundancy in the context surrounding the surface structure (that is, sentences coming before the one in question, or the individual's nonverbal behavior) to account for the deleted anchor.

For example, in the pair of sentences, "John ordered coffee and Mary ordered tea," and, "John ordered coffee and Mary, tea," the deletion of the verb "ordered" from the second clause of the second sentence is well-formed because there is sufficient redundancy in the form of the sentence structure to provide the needed information. However, such redundancy is missing in the sentence "John ordered coffee and Mary," or, "Mary ordered," unless the sentence is spoken by an individual who is holding some tea or pointing to some tea and using it to complete the communication.

Another type of deletion occurs in the sentence, "They ordered coffee and tea," where the individuals doing the ordering are deleted, unless they have been indicated elsewhere in the context. Lack of specificity, in this case, can constitute a form of deletion.

There are many types and levels of deletion. Many of the Meta Model strategies are geared toward being able to identify and retrieve deleted information.

2. *Simple Deletions* occur when there is a predicate (an anchor for a type of relationship between objects) in the surface structure, and one or more of the objects of the relationship are left out.

> Example: "I was always *told* that it was childish to cry." The programmer may respond, "Told by whom, specifically?"

Example: "I *feel* frustrated."
Response: "Frustrated by/with what specifically?" "Where do you feel frustrated?"

3. *Lack of Referential Index* is a type of deletion that occurs when an object or person referred to in the surface structure is left out or unspecified. Again, these objects or individuals may be indicated by context or nonverbal communication.

Example: *"They*'re always in my way."
Response: "Who/what specifically is always in your way?"

Example: *"That*'s not important."
Response: "What specifically isn't important to whom specifically?"

Example: *"It* always ends up *this* way."
Response: "What specifically always ends up what way?"

Note: Referential index is an important class of anchor, especially when the referent, "I," is involved. A programmer will be interested in determining when *symmetry* occurs with respect to referential indices and when a referential index *shift* occurs.

Symmetry occurs when the relationship between the individual and another referent is interchangeable.

Example: "You're not paying attention to me when/because you're not looking at me." "I'm not paying attention to you when/because I don't look at you."

Often the properties an individual ascribes to something else, or that she likes or dislikes in something else, will constitute properties that the individual wants, likes, or dislikes in his/herself. When an individual is talking about him/herself by assigning properties to other individuals or objects, it is called a *referential index shift*.

Example: *"I* want something *I* can cuddle and protect." "I want something to cuddle and protect me."

Example: "People are primarily motivated by guilt and sexual desire." *"I* am primarily motivated by guilt and sexual desire."

Example: "My mother doesn't respect herself." *"I* don't respect *myself* when *I* act like my mother."

Example: "My father thinks I should seek help." *'I think I should seek help."*

4. *Comparatives Deletion* is a special class of deletion that occurs when a referent of a predicate comparing two experiences (for example, good-better-best; more-less; most-least) is deleted.

> Example: "They did the *best* they could."
> Response: "The best for whom? The best compared to what?"
>
> Example: "She's the hardest person to understand."
> Response: "Hard for whom to understand?", "Hard compared to whom?"

5. *Unspecified Verb* is a class of words in natural language in which an anchor for some action does not supply all of the details of that action. *Feel, touch,* and *kiss* are examples of anchors in which the details of the action become increasingly more implicit within the experience anchored by the word.

> Example: "He really frustrates me."
> Response: "Frustrates you how specifically?"
>
> Example: "I tried to support her, but she rejected me."
> Response: "How, specifically, did you try to support her? How, specifically, did she reject you?"
>
> Example: "They hurt me very deeply."
> Response: "How did they hurt you, specifically?" "Give me an example of how you were hurt."
>
> Example: "She's just playing games with me."
> Response: "Describe to me an instance when she played games with you."

Clearly, verbs may be used to describe a very specific event, or a class of action. It is not always necessary for the programmer to get an explicit description of the client's internal experience. In fact, it is often more useful for the programmer to communicate ambiguously and assist the client in having more choices about an entire class of experience (hurt, for example).

The Meta Model responses are not used necessarily to challenge the client's way of organizing his/her experience (although this may sometimes be useful therapeutically), but to find out what the client's experience is; how he/she experiences hurt, frustration, play, and so forth.

6. *Nominalization* is a class of anchors that stand for actions or ongoing processes that, because of the position in the syntactic context, (as a noun), may distort the action into a static entity, deleting the objects or individuals responsible for the activity.

Compare the following sentences:
The carpenter built in the room.
The tension built in the room.

The programmer's response to the second sentence would most likely be: "Who's being tense about what, building how specifically?"

One of the simplest ways to identify a nominalization is to either imagine yourself holding on to the representation being anchored, or imagine putting it into a wheelbarrow. It is easier to imagine oneself reaching out and holding onto grapes than it is to imagine holding sadness or freedom. Similarly, it would be an easier task to put a carpenter into a wheelbarrow than tension.

When a programmer identifies a problematic nominalization, his/her most effective strategy is to restate the nominalization as the action it is indicating and recover the missing referential indices.

Example: "I can't stand her *insensitivity*."
Response: "Her not sensing what about whom? Sensing how specifically?"

Example: "They don't allow me the *freedom* I need."
Response: "You're being free to do what? Free how?"

Example: "This headache is ruining my *life*."
Response: "You living where, when, how, with whom?"

B. Limitations of an Individual's Model

An individual's *model* is composed of the anchors, and the ordering of those anchors, that an individual has developed to organize and guide his/her experience. Deletions and distortions of the types mentioned above may severely limit the programs an individual uses to organize his/her behavior. The next section of the Meta Model provides more specific classes of anchors that may limit an individual's repertoire of behavioral options.

1. *Presuppositions* occur when some experience must be implicitly assumed in order to understand the communicative sequence. Often, if taken for granted, these presuppositions limit or determine the scope of the communication. There are many types of presuppositions and syntactic environments for presuppositions. (For a listing, see the appendix in *The Structure of Magic,* Bandler/Grinder, 1975.)

> Example: *"Why* aren't you paying attention to me?"
> The presupposition is that you are not paying attention to me.
> Response: "How do you know that I'm not paying attention to you?"
>
> Example: "If you knew how much I suffered, you wouldn't act this way."
> There are three presuppositions in this statement: (1) I suffer, (2) you "act this way," and (3) you don't know how much I suffer.
> Response: (1) "How, specifically, are you suffering?" (2) "How, specifically, am I acting?" 3) "How do you know that I don't know?"

2. *Modal Operators of Possibility and Necessity* are the anchors that specifically identify the limitations of an individual's behavior. Modal operators of possi-

bility include words such as *can/can't, will/won't, may/may not, possible/impossible*. Modal operators of necessity include such words as *should/shouldn't, must/must not, necessary/unnecessary, have to/ don't have to*.

Again, the purpose of the programmer's response is not necessarily to challenge the individual, but to find out the individual's experience. Often the way a client experiences limitations internally will provide the programmer with very important information. Also important is the client's nonverbal behavior as he/she experiences a limitation (the way his/her eyes are looking when he/she forgets something, for example).

a. Possibility

Example: "I *can't* relax."
Response: "What stops you from relaxing?" "What would have to happen before you would be able to relax?" "How would you know if you were relaxed?"

Example: "I *can't* remember."
Response: "What stops you?" "How do you know you knew it in the first place?" "How would you know if you remembered it?" "If you could remember, what would you answer?"

Example: "It's *impossible* for him to change."
Response: "What stops him?" "How do you know that?" "If he could change, what would happen?"

b. Necessity

Example: "I *shouldn't* let anyone know how I really feel."
Response: "What would happen if you did?"

Example: "I *must* punish myself for breaking my agreement."
Response: "What would happen if you didn't?" "Have you ever not punished yourself for breaking an agreement?"

Example: "I *wouldn't* tell my mother what I truly think."
Response: "What would happen if you did?" "What stops you?" "Have you ever told her what you truly think?"

Neuro-Linguistic Programming

3. *Universal Quantifiers* are anchors that generalize a few experiences to a whole class of experience. Universal quantifiers are characterized by words such as *all, every, never, none, always.*

> Example: "She *never* listens to me."
> Response: "She *never* listens to you?" "Can you think of a time when she did listen to you?"
>
> Example: "I'm *always* uncomfortable around men/women."
> Response: "Can you think of a time when you weren't?" "Are you uncomfortable with me right now?" "What would happen if you weren't?"

4. *Complex Equivalence* occurs when the anchors for two experiences come to stand for each other, and yet the experiences are not necessarily synonymous. Often, there may be some causal connection implied between the two experiences in question.

> Example: "She's always yelling at me . . . she hates me."
> Response: "Does her yelling at you always mean she hates you?" "Have you ever yelled at anyone that you didn't hate?"
>
> Example: "When I'm rid of all my problems I'll be able to help other people."
> Response: "Can you help other people and still have problems of your own?" "Do you know anyone that has problems, but can help people?"
>
> Example: "He acts different . . . he's crazy."
> Response: "Can you think of what conditions would have to occur for someone to act like that and not be crazy?" "Would you be crazy if you acted like that?"

C. Semantic Ill-Formedness

Semantic ill-formedness is the section of the Meta Model that concentrates on the causal connections, or lack thereof, in an individual's ordering of his/her experience. The category, "complex equivalence," may also appear in this section.

1. *Cause-Effect* is a special class of unspecified verbs, particularly, *"make"* and *"cause."* Often a linkage is made between some external stimulus and the individual's internal experience or response, in which the two are not directly connected, or the connection is not clear. These often indicate unconscious synesthesia patterns like see-feel and hear-feel circuits.

> Example: "This *paper* makes me excited."
> Response: "How, specifically, does it *make* you excited?"
>
> Example: "Her insensitivity is *causing* the breakup of our family."
> Response: "How, specifically, does her insensitivity cause the family to break up?"
>
> Example: "I wish he wouldn't *make* me so nervous."
> Response: "How, specifically, is he causing you to be nervous?"

2. *Mind-Reading* tends to be a combination of presupposition, deletion, and referential index shift in such a way that an individual claims to know what another individual is thinking without having received any specific communications from that individual. These often indicate feel-see or feel-hear synesthesia patterns (the inverse of cause-effect).

> Example: "Henry never considers my feelings."
> Response: "How do you know Henry doesn't consider your feelings?"
>
> Example: "She doesn't understand the need for authority."
> Response: "How do you know that she doesn't understand?"
>
> Example: "You've lost your respect for tradition."
> Response: "How do you know that?"

3. *Lost Performative* is an anchor for a judgement or evaluation of some experience in which the context and conditions leading up to the evaluation have been deleted. Often a lost performative is a gen-

eralization or projection about "the world" that is based on the individual's own experiences. Lost performatives include such words as *good, bad, crazy, sick, right, wrong, true, false*.

> Example: "It's *bad* to argue."
> Response: "Bad for whom?" "Can you think of a situation in which it wouldn't be bad to argue?" "How do you know that it's bad to argue?"
>
> Example: "It's *useless* to try to understand what someone else is experiencing."
> Response: "Useless to whom?" "How do you know that it's useless?" "What stops it from being useful?"
>
> Example: "It's *wrong* to try to manipulate people."
> Response: "Wrong for whom?" "Can you think of a situation where it wouldn't be wrong to manipulate someone?"

It is evident that, in any verbal communicative sequence, there are many words and sentences that fit a Meta Model class, all of which cannot be challenged or elaborated. Also many verbal descriptions, taken in the context of the conversation, may be perfectly adequate descriptions of the client's experience. Further, there will be considerable overlap and combination of Meta Model categories. Often the programmer may wish to find out which Meta Model class the client uses most consistently, or which seems to be most problematic rather than immediately challenge an individual's surface structure.

3.4 Altered States of Consciousness: Accessing and Creating Parts

It was pointed out in the first section of this paper that a change in an individual's state of consciousness would involve an alteration of one or more of the parameters that make up such a state: (1) primary lead system, (2) primary representational system, (3) internal or external focus of attention, (4) hemispheric processing, and (5) amount of congruence between parts (parallel states of consciousness).

Sorting two simultaneously conflicting parts to be sequentially congruent, or anchoring two parts together, in essence would be leading the client into an altered state. The depth and intensity of an altered state that an individual may enter depends upon the scope and degree of consolidation of that individual's base line state of consciousness.

In NLP, altered states are used for three major processes:

1. *Sorting* simultaneously incongruent parts by creating states in which each part may express itself without interference from the other; that is, where the particular lead and primary representational systems' combination, comprising the part, has high signal values. This may be achieved by reanchoring aspects of the context in which the part was developed (age regression is one example of this).

2. *Accessing* different parts that the individual may not be able to anchor for him/herself in a typical state of consciousness. Multiple personalities are extreme examples of the results of such an inability. Addiction to alcohol, or other drugs that anchor an altered state for an individual, is another example. In such cases the programmer will want to assist the client in reaccessing the state in question (by controlling the context and supplying anchors from the "outside"), to give the individual an anchor for that state, and then to assist the individual in integrating that state with the client's typical conscious state.

3. *Creating* parts by reinforcing underdeveloped representational abilities or by integrating new representational combinations. Creating a "meta-part" that is composed of the integrated interaction of other parts is an example of this process. Often a part may be created or developed for task-specific purposes. Learning pain control may involve the development of a state in which the individual is disassociated from tactile sensations.

Learning to change one's state in order to learn some task, or class of tasks, is an example of Learning II (learning to learn). When the individual's task happens to be the ability to change states in response to tasks or contexts, the individual is learning Learning II. This type of learning is called Learning III. When a programmer uses techniques of altered states, all of these levels of learning are simultaneously involved.

It follows that anchors will be contributing to all of these processes as well. Anchors used in the formation of Learning I (learning as it is typically thought of) are called *conditioning stimuli*. These types of stimuli anchor a particular behavioral response or program (going when the light turns green, or adding numbers, for example). These anchors essentially constitute the content of behavior.

Learning II involves the change in behavioral content (conditioned responses) in response to varying contexts. In other words, Learning II is the form of behavior. The anchors that stimulate a change in the state of consciousness (the form of processing) in an individual are called *context markers*.

Conditioning stimuli and context markers are both made use of extensively in Neuro-Linguistic

Programming to assist the client in entering or developing a therapeutically altered state. There are three major strategies in NLP that a programmer may use to assist an individual into such a state: tracking, interruption of a behavioral pattern, and the exaggeration or enhancement of a behavioral pattern.

A. Tracking

The process of tracking an individual from one state to another involves the arrangement of anchors, by the programmer, so that he/she directs the client causally to a state of consciousness in which the client's representational hierarchy and configuration is different from that constituting his/her normal state or problematic state.

1. *Accessing the appropriate 6-tuples* is probably the simplest tracking technique. It follows the same principles as the three simple-anchoring techniques presented earlier in this paper. The basic premise here is that an individual can't not respond. Once the programmer identifies the state he/she wishes to direct the client to, he/she may use either standardized anchors (standard English vocabulary, for example) or anchors already developed personally by his/her interaction with the client to access 6-tuples appropriate to that state of consciousness. For example, the programmer may consistently use words such as "calm," "relaxed," "comfortable," etc. In order for the individual to make meaning out of these words he/she must access some degree of the 6-tuple experiences he/she has had for those states. The individual will access a different experience if the programmer mentions "agitation," "anger," "irritation," and so forth.

The programmer may simply emphasize these experiences in conversation or he/she may also ask the client to remember specific instances in which the client would probably be in the desired state. The types of anchors used by the programmer may be directed toward the state desired (for example, "Think of a time you felt really fatigued"), or it may indicate the state indirectly by anchoring a context in which the desired state would most likely occur (for example, "Have you ever stayed up all night writing a paper?").

Another consideration is the difference between anchors for behavioral content (Learning I conditioning) and those anchoring Learning II states. This would determine the difference between an individual's response to:

"Think of the color of your bedroom," or

"Think of what it's like to be in deep meditation."

Often a programmer can track an individual into an altered state by simply asking the client to discuss earlier altered state experiences that the client may have had.

6-tuples, of course, may also be anchored nonverbally with an appropriate tone or tempo of voice, type and tempo of gestures, breathing rates, type of clothes, and so on. Often, nonverbal anchors are the most effective Learning II state anchors. A tone of voice (as was pointed out earlier) tells a great deal about the way an individual is processing sensory information.

2. *Overlapping 6-tuples* is actually an extension of the previously mentioned technique. Overlapping is the process of accessing seemingly diverse 6-tuples that are unrelated excepting one or two shared com-

ponents. The effect of the overlapping tends to reinforce the shared components and increase their amount of signal. This technique is particularly useful for indirect or covert work. An example of this tactic would be to mention instances of reading very late at night, or instances of being sick as a child, or taking a nap after work. A probable point of overlap in these experiences is the act of lying in bed in a passive or weary state of mind. Experiences such as your first date, awaiting the dessert at a restaurant, or the night before your birthday as a child would share a component of anticipation.

3. *Pacing and leading* is one of the major meta-patterns of altered state strategies: that is, a pattern into which the techniques previously mentioned may fit.

a. *Pacing* is the process of recognizing the physical and representational components and parts that comprise an individual's state of consciousness at a certain point in time. This can be accomplished by using the identification procedures outlined earlier in this book and by increasing sensory skills and abilities to identify patterns of behavior. This information picked up by the programmer is then fed back to the client through the programmer's behavior. The feedback process may be accomplished through any set of sensory systems, but is most effective when it goes through that system which establishes the context markers (i.e., voice tone, body posture, physical contact, etc.) that anchor the state.

b. *Leading* is the process of guiding or changing the client's state of consciousness and behavioral patterns by changing the context markers established through pacing and initiating new anchors into the interaction.

Some examples of pacing and leading strategies are provided below:

1) *Matching predicates:* By listening to the predicates an individual uses to describe his/her experience, the programmer can tell a great deal about the nature of that experience and what representational schema the individual is using to organize that experience. It is, of course, important for the programmer to distinguish between those predicates that indicate and anchor experiential content, and those that indicate cognitive form.

> Example: Content: "His coat was *blue.*"
> State: "I couldn't *see* what he was saying."

> Example: Content: "Does that instrument have high *tonal qualities?*"
> State: "Does that name *ring a bell?*"

> Example: Content: "I like the feel of soft material."
> State: "He can't get a feel for basketball."

The programmer paces the client verbally by matching the class of predicates that the client uses to describe his/her experience and by using words and phrases he/she has heard the client use. When the programmer has successfully incorporated and reinforced these contextual properties of the client's state of consciousness, he/she may begin to systematically substitute predicates indicating representational properties different from those of the client's present state. Leading with content predicates will not necessarily appeal to or change the representational integrity of the client's state of consciousness, depending on the strength of and properties of that state (although it will in many cases). Content anchors appeal to an individual's lead system (for example, "Notice the *color* of the rug and the walls, then the *feelings* of your foot on the ground."). Changing lead system properties will change the client's state of *awareness*. Often, though, changing the context of a person's state of consciousness by varying the input signal (lead system–state of awareness) will lead directly to an altered state of consciousness.

Using state anchors directly can be very powerful, although they may often be interpreted and filtered through the client's existing state, which may reduce their impact.

> Example: "When you can really *get in touch with those feelings*, I want you to begin to *look at them very carefully* to see if they have any other *significance* for you."

The most powerful altered states will result from changes in both the content and the form of the client's experience.

2) *Body Mirroring* is a form of tracking an individual from one state to another by first pacing body movements, gestures, and postural qualities, and then introducing new body orientations into the interaction. The programmer may accomplish this by first imitating the client to some degree. It is important that the programmer not be too obvious, or contrived, during this process, or it may interfere with the pace. When the programmer can recognize the patterns in the client's body movements, his/her next step is to be able to initiate those responses in the client. This involves an understanding of the rules behind the client's body movements. Then the programmer may begin to vary the movements slightly, and observe whether the client follows. The

programmer may also *lead* by interjecting a new movement within a particular sequence of movements.

In this way the client may be led into new body positions and gestural sequences that may change the contextual properties of the interaction. As was pointed out earlier, body positioning, especially postural patterns, can be extremely effective anchors for different states of consciousness. Conditioning postural changes can also provide a powerful anchor that will carry over after the therapeutic interaction. (Tactile therapies, such as rolfing, massage, and other physical therapies, have many strategies for direct manipulation of body qualities by a therapist.)

3) *Pacing breathing* is another very effective form of tracking. Because of its effect on heart rate, circulation, metabolism, oxygen content in the blood, and so on, breathing is a very powerful contextual parameter in altered state technology. The programmer may progressively track changes in breathing rate by pacing and leading, as in the body mirroring technique described above, or he/she may direct the client's breathing rate verbally through suggestion, command, or by accessing 6-tuples that would probably have components, inducing an altered rate of breathing. (Rebirthing is an entire growth/therapy technique based on controlled breathing.) Because of its dual role, as an anchor and an accessing cue mechanism, breathing can anchor both content and state changes. Perhaps the most profound way to pace and lead breathing is with the tempo of your speech. Experienced hypnotists will almost invariably slow their tempo of speech and lower the pitch of their voice as they begin an induction.

The pace-lead strategy may be used to track *tonality, tempo, eye movements,* and any of the other accesring cues the programmer can identify. Often the programmer can lead his/her client's eye movements by directing them to a certain position with an arm gesture.

4) *Meta-pacing* is a tracking method in which the individual is paced for a particular behavior in a different sensory system than that in which the behavior is taking place. For instance, the programmer may verbally comment on the client's body movements rather than mirror them. Similarly, the programmer may pace the client's breathing rate with an up-and-down movement of his/her hand or with the sway of his/her body. The power of this type of pacing is that it introduces the activity of one representational system to that of another. This will often cause synesthesia or the overlap of different forms of experience.

> An example of pacing and leading via meta-commenting: "As you sit there feeling the chair against your body you know that I can see you breathing in . . . out . . . in . . . out . . . swallowing . . . blinking your eyes . . . swallowing again with your hands resting on your knees, your breathing rate and heart rate now beginning to change . . . and your eyes beginning to close *now.*"

4. *Linguistic tracking patterns* are patterns of ordering verbal anchors in such a way as to lead an individual away from their present state. In many ways, these patterns are used antithetically to those identified in the Meta Model. The programmer makes use of the Meta Model patterns to create new experience rather than specify old experience.

a. *Causal modeling* involves the establishment of cause-effect connections between two experiences that are not necessarily related. This occurs when the two experiences become complex equivalents, or anchors, for one another. This process is accomplished by connecting one anchor that acts as a pace with another that constitutes a lead. The connection is made by using a combination of modal operators of possibility and necessity and verbal anchors to indicate connection: that is, conjunctions, implied causitives, and causitives ("make" and "cause").

1) *Conjunctions*

"You can sit in that chair *and can* begin to really relax."
"You are listening to the sound of my voice *and* beginning to pay more and more attention to your internal dialogue."
"You are swallowing *and* starting to fixate your eyes on a certain spot."

2) *Implied causitives*

"As you listen to the noise of the cars outside you *will* feel more and more comfortable."
"The *longer* you stare at that object, the *harder* it will be to keep your eyes open."
"The *more* you try to move your hand, the *more* you'll realize that you *can't.*"
After you make that image clearer you *must* really be ready to enter a new state of consciousness."
"When my hand touches you, you *will* remember something that you have forgotten for a long time."

3) *Causitives*

"Closing your eyes will *make* you go into an even deeper trance."
"Taking deeper and deeper breaths will *cause* you to become more sleepy."
"Trying to keep your eyes open will *make* it more and more difficult."

In the surface structures listed above, two experiences are anchored together. One, the *pace,* is an *ongoing part of the client's behavior,* or some predicted pattern of behavior, which can become as effective an anchor for the *target experience,* the *lead,* as the words the programmer uses to initially access the experiences. Sitting in the chair is set up as an anchor for relaxation. If the behavior used as the pace is ongoing and repetitive (breathing or blinking, for example) the paired anchors can form a feedback loop for one another and become self-reinforcing.

b. *Lack of referential indices* and *unspecified verbs* can be very effective pacing-and-leading mechanisms. Because of their lack of specificity, they allow the client the maximum amount of flexibility and creativity in his/her response. In this way many resistances may also be bypassed.

> For example, the programmer may say, "As you sit all the way down in that chair you will *become aware of* that *certain sensation* getting stronger," or "When you listen to the sound of my voice and take a deeper breath in an effort to *become aware* of the comforts you have within yourself, you can *enjoy* every minute of *it.*

In the first sentence, "become aware of" constitutes an unspecified verb that allows the client the greatest amount of representational flexibility in awareness (as opposed to words like "feel," "see," "listen to," and so forth). The referential index, "that certain sensation," is also unspecific in this way. Anything the client becomes aware of, in any representational system, will constitute a successful response. The programmer may often use this type of verbal anchoring to establish an easy and consistent leading pattern with the client.

In the second sentence, the referential index "it" is unspecific as to which of the behaviors listed in the sentence it refers to. The client is left to choose what will anchor or be anchored by the experience of enjoyment.

"Become aware of" and "enjoy" are unspecified verbs that, again, allow the client the maximum scope of experience.

Similarly, a sentence such as, "You can really *learn* and *understand* the process of going into an even deeper state of trance" sets up the interaction so that the client can make the most use of his/her natural representational skills and abilities, because the way in which the "learning" and "understanding" is to take place is unspecified.

c. *Nominalizations* may be used in the same manner as the words identified above. A statement such as, "You can use these new *learnings* and *understandings* to change your *life*" uses three nominalizations (not to mention the unspecified verbs "use" and "change") that anchor very complex processes. This allows the client the maximum space to supply his/her own life experiences into the order implied by the sentence structure. The scope and complexity of the experience being anchored also forces a great deal of processing at the unconscious level.

d. *Presuppositions* involve the ordering of anchors for a context and a target response in such a way that the target behavior is already assumed.

> Example: "Do you want to go into a trance now, or five minutes from now?"
> Here, it is already assumed that the individual *will* go into a trance, the question is *when*.
>
> Example: "Don't go into a trance *too* quickly this time."
> Again, it is assumed that the person will go into a trance and that the person has been in a trance before, the only question is, how quickly.
>
> Example: "I wonder if you've noticed the chair that you will soon find yourself comfortably sitting in?"
> The presupposition is that you will sit in the chair.

Example: "I wonder if you're aware of how quickly you're relaxing?"

The presupposition is that you are relaxing.

e. *Modal operators of possibility and necessity, universal quantifiers,* and *mind reading* phrases may all be used in the process of introducing and accessing the appropriate altered state 6-tuples. These all tend to anchor and lead to Learning II states and relationships, rather than content.

Example: "You *(know, realize, understand)* that you *(can, will, must, have the ability to) (always, never)* pay attention to that certain experience *every* time I touch right here."

Example: "You *will* begin to get sleepy."

Example: "You *can't* get that certain thought out of your head."

Example: *"Everyone* goes into an altered state when they really listen to the sound of my voice."

Example: "After you come out of trance you *will never* have to worry about that problem again."

Example: "I know that *you are realizing* just how fast you *can* relax completely."

Example: "Your unconscious mind *knows* just when your hand *will* begin to lift."

f. *Lost performatives* and *comparatives* may also be used to introduce new anchors, contributing to the development of the altered state.

Example: "It would be *best* if you went into a trance as soon as you sat in that chair."

Example: "It's *true* that you will feel much *better* after you finish this trance."

Example: "It's *good* to really be able to relax."

Example: "The *right* thing to do would be to just sit back and make a clear image in your mind's eye."

g. *Tense* words are context markers that anchor a particular relationship between an individual's

Neuro-Linguistic Programming

6-tuple and his/her state. The present tense indicates the experience is here and now. For instance, compare:
1) You *were* seeing a blue dog in your head.
2) You *are* seeing a blue dog in your head.
3) You *will be* seeing a blue dog in your head.

In order to understand any of the above sentences, the reader must have, to some degree, a mental image of a blue dog: the anchor, "blue dog," is the same in all of the sentences. However, the past and future versions imply more attention to context than to the content being anchored. The programmer may use any of the various forms of tense, depending on the target state he/she has identified for the client. *Past tense* may be used for *dissociation* or the reintroduction of an anchor; *present tense* for vivification or *revivification; future tense* for the *establishment* of an anchor. Having a client use all present tense to describe some past experience can lead to a very profound altered state.

> Example: "You became sleepy as you gazed into my eyes."
> "You are becoming sleepy as you gaze into my eyes."
> "You will become sleepy when you gaze into my eyes."
>
> Example: "You felt very comfortable."
> "You are feeling very comfortable."
> "You will feel very comfortable."

h. *Lesser included structures* are words or phrases that are anchored, separately from the sentential context in which they occur, by nonverbal markings (that is, voice stress, gestures, eye movements, touches, and so forth). These markings may be used to emphasize some particular communication independently of the rest of the interchange, or to create an independent or embedded meaning within a seemingly ordinary expression.

Analogical marking is a covert tactic in which the programmer may use a change in tonality, posture, or tactile contact point (or other nonverbal cue) to mark out different parts of a seemingly continuous communication, as separate messages to two different parts of an individual.

> Example: "I told a client of mine the other day that *it is OK to be completely relaxed,* that she could *feel comfortable* and *really let go* around certain people, but that **you can be completely sensorily aware and awake whenever you need to be.**"

In the above example, the italicized words indicate the part of the communication that has been marked out as a particular message to one of the current client's parts, one that needs to be able to feel more comfortable. The phrase in boldface type indicates the marking out of a message reassuring another part that it will be able to take control, if necessary. The markings may be two different tonalities, two postures, a posture and a tonality, or whatever nonverbal anchors the programmer chooses that may have become associated with those parts during the client-programmer interchange.

Embedded commands are a class of lesser included structures that, using analogical markings, give a direct command to the client, covertly and indirectly. There are two major ways to deliver embedded commands:

a) Marking out the command as a subset of a particular sentence. This is typically done by embedding a command in a quote or a question:

> Example: "Somebody once said to me, *'your hand is beginning to lift without you noticing it.'*"

> Example: "I knew a man once who really understood that *you can really be happy if you put your mind to it.*"

Example: "I told the last person who was sitting in that chair to *take a deep breath and fall asleep.*"

Example: "I'm wondering whether *you can feel completely comfortable with me?*"

Example: "Do you know if you can *sit quietly and let your unconscious mind come out and talk to me?*"

b) Marking out the command over a number of sentences, by using some consistent, minimal cue to anchor bits and pieces of various sentences that can be put together to make a coherent message, when the entire communicative sequence is complete.

Example: "You can trust *your unconscious mind* to know that it *will* not have to *reveal to* your conscious mind anything that you don't want *me* to know, right now. *This uncomfortable incident* that you think is ruining your life is part of your past. You can come back *next week* and talk about anything you want, *in a direct and comfortable way.*"

In the above example, the programmer has marked out words, spread out over a number of sentences, that when put together make the embedded communications that "*your unconscious mind will reveal to me this uncomfortable incident next week in a direct and comfortable way.*" The minimal cue used to mark out the command (be it: touch, tonality, or gesture) will reanchor the simultaneously presented verbiage. The words, in turn, become the anchors for the embedded message.

This technique is, of course, very subtle and indirect, and the response the programmer gets may be equally subtle and indirect. It is up to the programmer to be keenly aware of all of his/her client's behavior.

i. *Derived meanings* are responses that are inferred from a particular communication. For instance, when someone says, "It isn't necessary to say anything," the inference is, "Don't talk." A negative, followed by a modal operator, often sets the stage for such derivations.

>Example: "There's no need to move."
>
>Example: "You don't have to listen to me."

Conversational postulates, which are in the form of questions, are another class of derived meanings. For example, when a person says, "Can you pass me the salt?", they are not asking a yes-or-no question, but requesting some action on your part. This same pattern may be used to direct an individual into an altered state.

>Example: "Can you relax?" "Will you be able to relax any more than that?"
>
>Example: "Can you put your hands in your lap?" "Are your eyes closed yet?"

j. *Suggestion and Suggestibility* have long been key concepts in hypnosis and other therapeutic uses of altered states of consciousness. It is evident from the preceding discussion that suggestions are verbal anchors for specified 6-tuples. As with all anchors, they are presented within a verbal and nonverbal contextual framework. The success of a particular suggestion is gauged by the client's response within that context. Likewise, an individual's "suggestibility" is determined in terms of that individual's response to anchors within a specified situation. In general, a person is not considered "suggestible" if they don't respond immediately to verbal anchors. In Neuro-Linguistic Programming, however, low suggestibility, or the inability to respond, may be considered a very powerful response.

Undoubtedly the verbal and nonverbal environment will determine the client's response to a great extent. The tonality used to deliver the suggestion will be as much of an anchor as the words. A demanding tonality, in a clinical situation, may increase the suggestability of one individual, but decrease it for another. A gentle or harsh tonality may anchor other 6-tuples that will overlap and interfere with an individual's response to the verbal portion of the communication.

Different syntactic styles (such as those listed in this section) will also make a difference in the client's response.

As was mentioned earlier, the client may be responding, but responding in a very subtle or indirect way. Often individuals will incorporate their response to some indirect suggestion into some ongoing behavior so that it appears to be a commonplace action rather than a response to a suggestion. Sometimes an individual will indicate they were consciously aware of a particular suggestion, but fail to realize that they are responding to it or have responded to it.

The polarity response is another common pattern, in which the client does the opposite of what he/she is instructed. For instance, they may be instructed to experience their arm becoming lighter, but instead experience it becoming heavier. This kind of behavior is considered resistance by some, but if the programmer notices a consistent pattern of response such as this, he/she may use it as effectively as if the person were responding "correctly" to the suggestions.

A client can't not respond to the anchors being provided by the programmer. It is up to the programmer to be aware of the nature and patterns of the client's response to successfully track him/her into an altered state.

The client's response to verbal tracking methods,

as with the other techniques, will be determined by the client's present state of consciousness, previous experience with altered states, sensory abilities, the kinds of 6-tuples the programmer chooses to elicit, and the context (that is, programmer-client relationship, syntactic styles, nonverbal minimal cues, and so on) in which the tracking takes place.

B. Interruption of Conditioned Behavioral Patterns

The quickest and most direct way to change an individual's usual means of processing information is to *interrupt* it. Interruption is one of the most powerful state-altering tactics available to the programmer. As such it should be used tactfully and carefully in some instances, for many people may experience the interruption of their meaning-making processes as abrasive, wrenching, frightening, and sometimes painful.

To understand the process more completely, refer to Appendix B for a discussion of some of the psychophysiological dynamics of the establishment of behavioral patterns.

1. *Blocking* a behavioral pattern is the process of, literally, not letting the client complete his/her typical behavioral program (as in interrupting a handshake). Examples of this include:

a) Interrupting the client in mid-sentence and/or changing the subject before the client has completed what he/she was going to say.

b) Acting inappropriately to the context.

> Examples: Bizarre behavior, such as putting your coat on backwards.

Neuro-Linguistic Programming

> Talking in word-salad, or using words or phrases that don't make sense.
>
> Suddenly manifesting the same symptoms as the client.
>
> Encouraging the client's symptoms.

c) Responding inappropriately to another's communications.

> Examples: Laughing at the wrong times.
>
> Consistently responding incongruently.
>
> Interjecting a phrase or gesture inappropriate to the interchange.

d) Blocking accessing cues.

> Examples: Moving your hand(s) rapidly in an individual's visual field to interrupt lateral eye movements.
>
> Making an individual breathe consistently from only one part of their body and at a certain tempo.
>
> Tactically interrupting an individual any time he/she tries to access.

e) Forcing an individual to be conscious of typically unconscious, consistent behaviors.

> Examples: Giving the individual feedback for breathing, lateral eye movements, consistent gestures, predicates, etc.
>
> Meta-commenting on all of the client's observable behavior.

f) Giving suggestions to the client as he/she is talking to you: talking at the same time as he/she are.

g) Have two people talking to the client at the same time (a double induction).

h) Unexpected or unfamiliar stimuli.

A beautiful example of pattern interruption is described by Milton H. Erickson in *Advanced Techniques of Hypnosis and Therapy.* He states:

"One of the physicians present (at a lecture on hypnosis) was most interested in learning hypnosis, listened attentively during the lecture, but in the social hour preceding the lecture, he had repeatedly manifested hostile aggressive behavior toward most of his colleagues. When introduced to the author, he shook hands with a bone-crushing grip, almost jerked the author off his balance (the man was at least 6 inches taller than the author and about 65 lbs. heavier) and aggressively declared without any preamble that he would like to 'see any damn fool try to hypnotize me.'

When volunteers for a demonstration were requested, he came striding up and in a booming voice announced, "Well, I'm going to show everybody that you can't hypnotize me." As the man stepped up on the platform, the author slowly arose from his chair as if to greet him with a handshake. As the volunteer stretched forth his hand prepared to give the author another bone-crushing handshake, the author bent over and tied his own shoe strings slowly, elaborately and left the man standing helplessly with his arm outstretched. Bewildered, confused, completely taken aback at the author's nonpertinent behavior, at a total loss for something to do, the man was completely vulnerable to the first comprehensible communication *fitting to the situation* that was offered him. As the second shoe string was being tied, the author said, "Just take a deep breath, sit down in that chair, close your eyes, and go deeply into a trance." After a brief casual startled reaction, my subject said, "Well I'll be damned! But how? Now do it again so I can know how you are doing it."

He was offered a choice of several traditional techniques. He chose the hand-levitation method as seeming the more interesting, and this technique was employed slowly both for his benefit and that of the audience, with another somnambulistic trance resulting."

2. *Confusion techniques* incorporate many of the tactics above and also involve the use of *ambiguity* by the programmer. Ambiguous anchors are those that have multiple meanings and call for a number of different responses from the client. Confusion techniques involve supplying a syntactic environment such that the appropriate response to the anchor cannot be derived from the context.

Grinder and Bandler (1975), identify a number of different types of verbal ambiguities:

a. *Phonological ambiguity* involves the overlap of words that sound alike but have different meanings (like *see* and *sea*).

> Example: "Is your hand *apart* or *a part* of the feelings of the *rest* of your arm?"
>
> Example: "If you were carrying a bag of rocks in the darkness, how could you make your journey *lighter?*"

Phonological ambiguities may also be used to deliver embedded anchors.

> Example: "When you *get* to the *well* in your dream, bathe in its healing waters."

Some other phonological ambiguities include: *weight/wait, read/red, hear/here, and knows/nose*.

Bandler and Grinder point out that "one very rich source of these word ambiguities is pairs of words that are ambiguous with respect to their syntactic category. Many verb/nominalized verb combinations have this feature: *lift, pull, point, touch, push, nod, move, talk, shake, hand, feel*."

b. *Syntactic ambiguity* consists of anchors in an environment so that the function of the word cannot be uniquely determined by the rest of the sentence.

> Example: *Hypnotizing* hypnotists can be tricky.
>
> Example: Imagine the *feelings* of a guinea pig in your one hand turning into the *feelings* of another hand.
>
> Example: *Feeling* hands can be a new experience.

The pattern to generate such ambiguities is Verb + ing + Noun.

c. *Punctuation ambiguity* occurs where two sentences overlap where they share the same word.

Example: "Can you tell me the time on your *watch* my hand carefully."

Example: "I want you to notice your *hand* me the glass."

Example: "That's a good *point* to the place that's causing you trouble."

The multiple or incongruent responses, anchored by the use of ambiguity, cause a state similar to the interrupted state that may be used to reprogram or deprogram unconscious patterns of behavior. The most powerful confusion techniques incorporate both ambiguity and interruption. For instance, the programmer may suddenly stop in the middle of a sentence, reach over, and ceremoniously touch the subject in a number of places, then stare at him/her directly in the eye, and say meaningfully, "Which time were you touched more times than the time before you were last touched?"

All the programmer will need to do later to reanchor the state will be to look the subject directly in the eyes, stop mid-sentence or ceremoniously touch the client.

Interruption techniques are the most effective techniques for reprogramming patterns that are very deeply ingrained, or for difficult situations where gentle suggestions do not seem to have much effect. I have found pattern-interruption techniques most effective in dealing with "psychotic" patients (who, incidentally, are very good at interrupting other people's typical programs).

C. Enhancement and Exaggeration of a Behavioral Pattern

A controlled reinforcement or repetition of some pattern of behavior can be an extremely profound means to alter an individual's state of consciousness or alter the 6-tuple/program anchored by some stimulus. Biofeedback is an example of this, particularly alpha-wave feedback, where an anchor is established for the alpha state through a tone.

The controlled repetition of some behavior tends to change the signal value of different parameters of the 6-tuple. Habituation and desensitization occur for some parameters, while others may become conscious for the first time or become hypersensitive. A word said over and over again tends to lose its meaning. Chants and mantras are common examples of this process used to develop altered states of consciousness. Fritz Perls made use of this process in his work by asking clients to repeat key phrases repeatedly, forcing them to be conscious of their tonalities and language processes.

Exaggerating accessing cues and self-established anchors (such as gestures and utterances) will often lead to altered states of consciousness, because the process intensifies the states that it anchors. When an individual exaggerates a response to such an extent that it becomes an entire state. The individual is said to have "*Lost Quotes*" in that response.

Intensifying a behavioral pattern to create a state of consciousness will reaccess 6-tuples and portions of 6-tuples unique to that state, often those 6-tuples contributing to the development of the pattern. (When something goes wrong it is easier to think of all the times things have gone wrong than when they've succeeded.) The process of accessing 6-tuples through the intensification of a state is known as a *transderivational search*.

This process can be used very effectively to reframe resistances in a client. If a client is having difficulty responding, the programmer, by observing his/her behavior, may notice some behavior (perhaps very minimal) that seems to coincide with his/her failure to respond. The programmer may then stop the interchange, and ask the client to intensify this behavior and report what he/she is experiencing in his/her various sensory modalities. Some clients may, for instance, see their mother in their mind's eye, ordering them to do something, when they were adolescents. By having the client intensify the behavior even more, and allowing the 6-tuple to change, he/she may regress even further in age, or access a more vivid or intense 6-tuple. By then reprogramming these older responses, the programmer can track the client to new responsiveness in the present situation.

Sometimes the exaggeration of one particular state/part will anchor a state that is completely different. When the part has characteristics opposite of the generating state, it is called a *polarity flip*. The flip will most likely occur as a result of the transderivational process.

3.5 Metaphor

Metaphor is probably the most elegant tool, for assisting a client in changing, available to the programmer. Metaphor is defined as *"a figure of speech in which something is spoken of as if it were another."* The development of the *"as if"* phenomenon involves two major components: *isomorphism* and *symbolism*.

Symbolism involves the establishment or displacement of a referential index. A symbol is a character, object, or situation that embodies (becomes an anchor for) certain behavioral responses. The programmer will, of course, want to establish symbols for responses that pace and eventually lead those responses within the client's present repertoire. For example, the programmer may give the client an indirect message or suggestion prefaced by, "I had a client once. . . ." The fact that the individual being talked about and the individual being talked to share the distinction of being "clients" establishes the possibility of similarity of response.

Isomorphism involves the formal similarities between representations of different responses. Oscillation, for instance, may be an identifiable property within a number of varied systems from electrical circuits to a spring, to guitar strings to the relationship between two individuals. Individuals can learn much about the possibilities of their own behavior by considering the operation of other systems. Imagining that you are a bird in a certain situation, as opposed to a lion, will open up and abolish many different avenues of response. *Cybernetics* is the field that studies the formal similarities between different systems. (See Part I of this work.)

In general, symbols will identify the structural aspects of the metaphor, while isomorphisms will deal with the relational or syntactic components.

By establishing an anchor through some character in a story (whether it is an animal, a rose, a queen, or another client) for some behavior that paces that of the client, a chain of events can be formulated through a story line or plot in which new responses may be generated or suggested, or established responses may be integrated or changed by the client internally via the interactions of the symbol/anchors. The programmer may supply the whole progression of events, or allow the client to participate by filling in responses, events, and characters (which the programmer may use to gather information indirectly about the client's internal processes or experiences).

Setting up a story and allowing a client to solve his/her problems metaphorically is an extremely effective tactic.

Embedded messages, analogic markings, and all of the other linguistic patterns listed in this paper may also be employed within the context of a metaphor to enhance and increase the effectiveness of its outcome.

An example of using a metaphor for a part's integration is given in Appendix A of this paper.

A. Isomorphism within Neuropsychological Systems

Of extreme importance to the success of the metaphoric process is the ability of neural networks to incorporate information about classes of behavior. This is what makes Learning II (learning to learn) possible. This ability is perhaps most exemplified in an animal's ability for "set learning" or to become "test-wise." An animal trained in instrumental avoidance behavior will be able to learn new and different types of instrumental avoidance more and more rapidly. It will, however, tend to be slower at learning some Pavlovian-conditioned behavior than an animal that has been conditioned in that class of behavior earlier. It has picked up the rules or patterns for a whole class of behavior.

Traits of personality and personal abilities are undoubtedly established through isomorphisms in behavior. Individuals confronted with new situations will respond to them in a manner analogous to contexts they have encountered before. We may find that an individual's eating habits are analogous to his/her sexual habits, which are analogous to the way he/she talks, plays the piano, or interacts with strangers. In a sense, anything that an individual does will be a metaphor for the way that he/she organizes his/her experience. The way that an individual talks about neural interactions will reflect his/her own thinking processes. Representational system primacy is another example of the development and establishment of classes of neuropsychological operation.

Gregory Bateson calls the process of predicting isomorphisms in an individual's behavior *abduction* (as opposed to induction or deduction). The well-known phenomenon of *transference,* where one individual responds to another as if he/she was some "significant other," is another product of behavioral isomorphisms.

Milton H. Erickson gives an account of how he used this process while working with a couple having marital difficulties over their sexual behavior. Erickson talked to the couple about their eating habits. He found that their eating habits paralleled the individual sexual behaviors that were causing the difficulty. The husband was a meat-and-potatoes man and liked to head right for the main course, while the wife liked to linger over appetizers and delicacies. For their therapy, Erickson had them plan a meal together in which they both were able to attain satisfaction. The couple, of course, had no idea of the significance of the event, but were pleasantly surprised to find that their sex life improved dramatically afterwards.

Isomorphisms in behavior stem from two basic psychophysiological phenomena.

1. *Behavioral generalization* occurs when an animal responds to one stimulus as if it were another: when a cat responds to a light flashing at 10 Hz with a response that was originally conditioned at 5 Hz. E. R. John writes:

> When a trained animal performs a conditioned response to a novel stimulus, indicating behavioral generalization, some brain regions display EEG rhythms corresponding to the frequency of the conditioning stimulus actually used in training rather than to the frequency of the novel stimulus which elicits the behavior.

Generalization may occur for a number of parameters and can take place cross-modally, as when a trained animal responds to a 10 Hz auditory rhythm with a behavior originally conditioned for a 10 Hz visual cue. The type and amount of generalization will, of course, depend on the discriminative capabilities developed by the individual animal, which will be affected by the various levels of development of the individual's lead and primary representational systems.

The context in which the behavior is taking place will provide other anchors that will also have a great effect on an animal's discriminative efforts. The development of "experimental neuroses" in laboratory animals is a good example of this.

A dog is trained to discriminate between two stimuli, an ellipse and a circle, for example. It is then made to continue making the discrimination as the two stimuli are made to match each other more and more closely (the circle is flattened and the ellipse is fattened) until discrimination becomes impossible. At this point the animal will begin to exhibit symptoms such as refusal to eat, attacking the trainer, even becoming comatose. A naive animal that has not been pre-trained, when presented with the indiscriminable stimuli, will not show any of these symptoms, but simply guess randomly. For the trained animal, the class of responses that should take place (the representation of which is anchored by the laboratory environment) is incongruent with the behaviors available to the animal.

The above example also provides a good deal of insight into the etiology of much neurotic and psychotic behavior and the role of parts.

The programmer should keep this in mind as he/she sets up the context of interaction with his/her client. The context established by a metaphor may encourage behavioral generalization for a number of parameters.

2. There is evidence to show that the cortex in the brains of higher mammals stores and processes information about what might be called "relational abstractions" as well as simply providing a map for perceptual and motor details. That is, the cortex processes information about classes of relations as well as the details of the relations. Pribram (1976) for instance argues that *actions,* not just *movements* or *muscles,* are represented in the motor cortex. An example of this is that, without ever having even tried it before, most people could write a pretty reasonable facsimile of their name with a pencil held between their toes. The action, "signature," has been stored above and beyond the movements of the muscles we used to originally learn and generate the action.

Pribram conducted experiments involving the removal of various amounts of motor cortex in monkeys. He reports:

> Even extensive removals failed to paralyze any particular muscle or muscle groups. Nor did cinematographic analysis show any specific movement (sequence of muscular contractions) or sequence of movements to be disrupted by the ablations. Yet skill in certain tasks was impaired (latencies for completion of latch box puzzles became prolonged). I interpreted these results to mean that neither muscles nor movements were represented as such in the cortex—that instead, actions, the specific environmental outcomes of movements, were represented.
> ... [Evarts, 1967] showed that neurons in the motor cortex of monkeys do not fire proportionately to the amount of lengthening or shortening of a muscle involved in depressing a lever. Instead firing is proportional to the weight attached to the lever, i.e., the force necessary to move the lever. It is not the muscle or its contraction, it is the act, the use to which the muscle is put, the predicted ends that need to be achieved, that is reflected in the activity of the cortical cells.

One consequence of this is that it allows for considerable freedom and plasticity in the composition of a response or program. By extending the pattern to the other parameters of the 6-tuple, we may also predict the ability for considerable transference and overlap of responses.

There will always, of course, be structural limitations to the isomorphism: There will always be a point at which the isomorphism breaks down. It would be practically impossible for an individual to write their name with his/her earlobe, for example.

Further, the programmer will have to deal with any structural-functional dependencies developed for a particular class of responses—either hemispherically or within a particular representational system. (Such dependencies will be contributed to by the phenomenon of lateral inhibition within neural networks.) Interruption techniques will be the most effective for these.

IV. Therapeutic Epistemology

The purpose of this section is to provide a reiteration and outline of the fundamental premises of Neuro-Linguistic Programming and their implications for the field of psychotherapy and professional communication.

4.1 All behavior is the result of biological interactions; "mind" and "body" being part of the same biological systems. In human beings behavior is initiated and modulated via neural networks consisting of peripheral receptors, central-processing and integrating networks in the brain, and motor effectors through which information is output to generate observable behavior. The receptive perceptual systems are responsible for inputting different classes of distinctions about the individual's environment, and as such, will initiate different classes of motor transforms as output. The most important perceptual input/motor output complexes, or representational systems, are auditory, visual, tactile kinesthetic (somatosensory), and visceral kinesthetic (limbic).

Behavior is defined as activity within any of the representational system complexes, and ranges from seeing in the mind's eye, internal dialogue and body sensations, to lateral eye movements, breathing rates and muscle twitches, to language, gestures, and art.

The sensory-motor complexes making up the various representational systems provide the structural components of behavior. The dynamics of behavior, behavioral programs, are the result of interactions between structural entities: resonant and desynchronized interchanges involving the various sensory-motor networks that are recordable as varying electric potentials. Consistent and recurrent behaviors are established, due to the resonant properties of structural elements and changes made in the synaptic and dendritic connections of brain cells. A particular program of behavior is consolidated as a common mode of activity within extensive neuronal populations that is recordable as a synchronized brain rhythm within those areas.

This activity is dispersed through the various representational systems with varying signal densities. A particular *6-tuple* of experience (a term used to describe the various types and amounts of activity within the various representational systems at any particular point in time) is used to describe the cognitive aspects of a program. The establishment of a program is such that any parameter of the participant representational systems in the *6-tuple* (A, V, K_t, K_v, O, M) may initiate and spread the common mode of activity necessary to reaccess the program of behavior to some degree. The initiating stimulus is called an *anchor*.

As a pattern of behavior becomes more established and habitual, the amount of synchronization between the participant representational systems allows the respective signal values in those systems to drop, making the activity less and less "conscious."

4.2 *Representational system primacy* (the behavioral dependence on one mode of perceptual information over the others) evolves from the hypertrophy and/or atrophy of neural systems and interactions due to internal and external environmental conditions as an individual matures. The evolution has to do with the effects of sensory stimulation on cell migration during maturation and the effects that changes, made neurologically during the incorporation of previously established programs, have on the codification and consolidation of new programs (Learning II).

These properties lead to the development of isomorphisms between the behavioral programs of an individual, elicited in response to completely different contextual conditions (for example, eating and making love). This is probably a result of the types of resonances that make neural synchronization possible, and the structure of cortical elements.

They also allow for a certain amount of plasticity in the elements involved in achieving a behavioral outcome.

Lead system primacy evolves in a manner similar to representational system primacy, but has a different functional significance. A lead system is the representational system that becomes most proficient at anchoring (accessing or recalling) 6-tuples of which it was a participant. It becomes (due to structural development and synaptic changes) the prime initiator of the common mode of activity that leads to synchronization identifying a particular behavioral program. The activity in the lead system may be generated externally, through receptors, or generated internally, through activity in some other representational system.

The difference between a lead system and primary representational system is that the *primary repre-*

sentational system supplies the distinctions and processing patterns most significant to the behavior that is eventually output; the *lead system* initiates the common mode of activity that supplies the "juice," so to speak, for the response. In general, the lead system determines the *affective* distinctions, and the primary representational system determines the *effective* distinctions.

The therapeutic importance of lead and representational systems is obvious: (a) they will influence and limit the distinctions used to generate responses to an individual's environment. They will define which portions of the universe and which classes of distinctions from which an individual will draw to establish programs of behavior. The particular combinations of lead and representational systems an individual uses to establish behavioral programs is called his/her *learning strategy*. (b) They will establish certain isomorphisms in the client's behavior that will allow the therapist to predict certain classes of response within the client's behavior.

4.3 *Parts* of a person result from changes in the functional significance of the various representational systems as a result of contextual conditions. A part develops when some behavioral pattern is reinforced and consolidated to the extent that the synchronization of neural activity sets up a common mode of firing that is strong enough, and occupies extensive enough neuronal populations, to affect the individual's entire brain state. The coherence of the state causes enough change at the neurological level to influence the establishment of new programs and to create properties of isomorphism consistent with the synchronization of neural activity defining the part/state.

Because of the consistency in synchrony, the programs anchored in the state/part will be unique to that state. The part, then, establishes a persona defined by its component lead-representational system/synchrony properties.

Due to structural properties of the brain (like the existence of the two cerebral hemispheres), parts may become established, not only sequentially in response to varying contextual conditions, but concurrently when there is enough redundancy in the context (both internal and external) to access both states. Depending on the properties of the situation, simultaneously elicited parts may (a) vie for behavioral control until one dominates, (b) integrate, creating a third behavioral response that is a combination of the two, or (c) manage to coexist by controlling functional domains that do not overlap.

4.4 *Mental health* in NLP is considered to be the ability to have access to all representational abilities so that the individual will be able to respond appropriately and effectively to a variety of contexts and environments. (This, of course, does not imply that all individuals should respond with the same behavior in the same contextual setting, as different responses will be adaptive for different individuals in different ways.) This requires the development of a number of different parts and implies a certain degree of harmony between the parts. Conflict between responses will impair the effectiveness of the behavioral outcome. Further, due to the nature of the neural interface at which these conflicts take place, the results are not limited to cognitive activity within participating representational systems, but affect output to all regions of the body and may cause physiologic pathologies in any number of areas.

4.5   The human biological system is presupposed to be adaptive. That is, it is supposed that it will seek a steady state unless there is serious insult or damage to the system. Maladaptive behavior and conflicting multiple responses are not, then, necessarily, considered to be the result of psychological pathologies. In fact, it is an underlying presupposition in the therapeutic epistemology of NLP that *all behavior is, or was, adaptive for the organism, given the context in which it was learned,* and that *all individuals make the best choices available to them,* given the representational abilities and behavioral options available to them. Psychotic, criminal, and neurotic behavior are all considered to be the best choices of behavior available to the individual exhibiting them, at the point in time of the occurrence.

The goal of NLP is to give the individual confronted with problematic behavior more choices in their repertoire of response through sorting, accessing, creating, and integrating parts. This requires the therapist to be a true programmer: a behavioral technician constantly involved in the restructuring and reordering of behavioral responses with regard to changes in context.

4.6   One of the most important and powerful tools available to the programmer is his/her ability to be completely aware sensorily of all the perceivable behavior offered to him/her by the client. The human biological system is cybernetic. Therefore, it is accepted that (a) *any occurrence in one part of the system will necessarily affect all of the other parts in some way,* and that when the rules of interaction between the parts of the system are understood, the effects of the different parts of the system on one another can be patterned, predicted, and changed. The programmer's ability to use his/her sensory abilities to find the patterns, or regularities, between an individual's overt behavior and his/her internal processes is invaluable: (b) *all behavior* (from pigment change to predicates) *is in some way a transform of internal neural processes,* and therefore carries information about those processes. *All behavior is communication.*

The programmer uses his/her sensory abilities and the principles and techniques of anchoring (outlined in this paper) to create and integrate new choices of behavioral response for the client.

The technique of *pacing-and-leading* involves the programmer's ability to abduct isomorphisms in the client's behavior to predict his/her behavioral responses to certain classes of stimuli, and then either reinforce or interrupt those patterns of response.

4.7 Change occurs as an alteration in the dynamics of an individual's 6-tuple/program of response to some stimulus. Because programs are the result of internally generated synchronized neural activity, it is obvious that the experience causing the change need not occur in some "external" consensual reality. It should also be evident that the individual need not be conscious of the stimulus that creates the change, nor the process of change. What is important is an alteration in the syntax of endogenous programs. The electrical potentials needed to produce this alteration may be as easily, and strongly, generated internally. (Incidentally, the cause of the change need not be conscious to the programmer either.)

4.8 It is important for the programmer to be able

to make the distinction between classes of behavior that constitute a state (Learning II processes) and the programs that make up that state (Learning I). There tend to be different schedules of reinforcement for a behavior that is subsumed under another one. Gregory Bateson supplies a good example of this difference in *Steps to an Ecology of Mind:*

> ... You can reinforce a rat (positively or negatively) when he investigates a particular strange object, and he will appropriately learn to approach it or avoid it. But the very purpose of *exploration* is to get information about which objects should be approached or avoided. The discovery that a given object is dangerous is therefore a success in the business of getting information. The success will not discourage the rat from future exploration of strange objects. (p. 282)

The more acute behaviors, such as bad habits, compulsions, and phobias tend to constitute content behaviors and are fairly easily dealt with by employing simple anchoring techniques of deprogramming and program substitution. Behaviors such as chronic depression, psychosis, or neurosis will probably require state-altering techniques, such as interruption, exaggeration or the various verbal and non-verbal tracking techniques.

There is, of course, a good deal of overlap between the two classes of behavior, and a good deal can be done with chronic illnesses (including physical ailments like tumors, arthritis, pain, and so on) in an hour's work.

4.9 Being an effective programmer essentially requires that the therapist have enough *sensory experience* to get *feedback* for the way that he/she is manipulating the context (internally and externally) that is determining the client's responses. A good programmer must then have enough *flexibility of behavior* to continually alter his/her own behavior until the *desired responses* are elicited in the client. There is no such thing as resistant clients, there are only inflexible therapists.

Appendix A

INSTANCE OF INTEGRATION OF
PARTS USING REFRAMING AND
METAPHOR

During an exercise in the course of a training session held by the author, R, a licensed counselor, complained of experiencing bothersome lapses in attention during the group proceedings. The content of her experience during these lapses in attention was unavailable to her at the conscious level. She felt, however, that they caused her to miss a good deal of the material being presented.

She linked this problem to two other consistent experiences which had recently become more and more pronounced each day: 1) increasing difficulty in getting out of bed in the morning, and 2) a painful anxiety in the evenings after work concerning her abilities as a mother, wife, and therapist.

It was noted by the author that R's body was slender and evenly proportioned. Her flesh and musculature was soft and relaxed. She did not wear glasses or contact lenses and had rather large prominent eyes. Her overall tonality tended to be low and soft with a relatively slow tempo and noticeable pauses between words. This gave the author the impression that R had strong visual (slender body and prominent eyes) and kinesthetic internal (soft flesh and muscles and low tonality and slow tempo) parts. Due to the even distribution of her body weight it was hypothesized that any conflict of these parts would probably take place sequentially (each expressing itself independently of the other) rather than simultaneously. (Individuals whose visual and kinesthetic internal parts tend to express themselves simultaneously will tend to have a pear-shaped body; thinner and tighter at the shoulders, neck, and chest, and soft and fleshy in the lower stomach, hips, and thighs.)

R was then asked to elaborate on the two patterns of behavior that she had identified as becoming increasingly problematic. It was noted that, as she communicated, R was consistently symmetrical with hand gestures, constantly moving both hands in unison, except when she was talking directly about her experience during one of the two identified states. When she talked about her difficulty getting out of bed in the morning she would invariably gesture with her right hand, the other resting immobile in her lap. As she talked about her anxieties she would consistently gesture with her left hand; the right hand, in this case, remaining immobile.

Additionally, it was observed that as she commented on her feelings of anxiety R's tonality would become higher and more strained in pitch and that tonal shifts of this type were consistently preceded by lateral eye movement up and to the right and a drop or even a pause in her breathing rate. As she discussed the other behavior, however, it was observed that her tonality dropped noticeably to a much lower and softer pitch, and that her tempo became much slower with long pauses between words accompanied by the taking of a deep breath or a sigh. This was preceded by head and eye orientation down and left.

As R talked, each state was anchored kinesthetically by a slight touch of the author's hand on the knee of R that corresponded to the observed asymmetry in R's hand gestures. The contextually based changes in R's behavior were then commented on to the other members of the group, who verified the observations. To the aforementioned observations, R added that she was left-handed.

The content of R's comments on her experience was as follows: A) Her inability to rise from bed was experienced as a feeling of "heaviness" all through her body and "a feeling of dread at the prospect of having to move or do anything;" B) The anxiety was experienced primarily as a kinesthetic agitation in her chest area, which she termed painful, accompanied by an internal dialogue in which she berated herself for being lazy and not doing as much she could to improve her skills and abilities.

The process of reframing the behavior was then begun and R was instructed to go inside and contact the part of herself that was responsible for each of these experiences, in whatever way was most effective for her, and to ask those parts what they were trying to do for her. In this way it was hoped to separate the adaptive intent of the behavior from R's negative experience of it.

As R contacted each part, the author reanchored the experiences associated with that part by touching the appropriate knee. Contact with the part was verified through the observation of the appropriate breathing, gestural, tonal, and eye position changes.

R responded by saying that the part characterized by the heaviness and dread of action had developed over the course of her 11-year marriage. She felt that it may have been a response she evolved to keep herself in the home and not feel bad about staying with the children.

Her children were just now becoming old enough that she could leave them alone and she had just recently resumed practice as a counselor. She commented that the anxiety had been getting stronger since that point. She said that the anxiety had kept her from getting overly sedentary and festering but that it was "so painful." She thought that the part had been overcompensating. Ever since she had resumed practice she had been afraid to stop learning and preparing herself. She felt compelled to attend workshops even though a part of her knew that she had enough preparation.

She was then asked to access this part of her by telling the author about a client she had worked successfully with. Her voice immediately picked up clarity and volume, and her mouth held a constant symmetrical smile. As she related the incident she folded her hands and held them just below her chin, just out of contact with her chest.

She was then asked to relate an experience in which she felt integrated, relaxed, and comfortable. R described an incident that took place while she was on vacation in a low but sure voice with her hands resting, folded in her lap.

After this series of responses was elicited and each 6-tuple anchored respectively, the author planned to do a kinesthetic integration (by touching off the appropriate anchors for the two conflicting parts simultaneously) as a demonstration. R was asked to reaccess the two parts in question but experienced difficulty "getting in touch with them." She complained of having a hard time "holding on to" the feelings, that they weren't "clear" and were difficult to deal with.

After listening to her descriptions, the author postulated that it might be advantageous to make use of R's well-developed visual system to establish a "clearer" anchor for the parts in question. She was instructed to hold out her right (non-dominant) hand, palm up, orient her head in that direction, and "make a picture of that part."

A look of surprise momentarily passed over R's face and then she reported, "It's me . . . but much heavier. I can hardly recognize myself. She must weigh about three hundred pounds." She continued her report in a low, slow tonality, "She's soft . . . sort of sluggish . . . very comfortable in a sensual way, not necessarily happy . . . She doesn't look ahead to anything. She doesn't want to move because she's comfortable where she is."

Her eyes had closed during the description. When she finished she was asked to, without opening them, look over at the hand and see what that part looked like.

R's voice picked up. "She has fiery eyes and is always moving . . . She's much more aggressive . . . and has a real need to learn new things . . . she wants to keep moving ahead."

As R talked there was much more visible movement in her body.

The body type and behavioral characteristics of the first part she had pictured seemed to characterize perfectly those of an individual whose primary lead and representational system were almost exclusively visceral-kinesthetic. The other, with fiery eyes, seemed to be a part that led visually (with an external orientation) and represented the information by doing, utilizing the motor and tactile systems.

R was then asked to look in between the two hands

and see herself as therapist. The therapist's physical appearance was in between that of the other parts, and seems to incorporate the best characteristics of both.

R was then instructed to have the parts look at each other. She cocked her head slightly to the left and then right, furrowing her eyebrows slightly, as she momentarily assumed the identity of each.

She said that the two looked very strange to each other and that they had never really seen each other before.

The author then asked R if she told stories to her children. A rush of color came into her face and she smiled and responded, "yes," and said that she really enjoyed it. She was then asked if she could have one of the two parts tell the other a story about two people just like them who were able to learn to live together in harmony. It was added that she need not do it out loud.

She closed her eyes, cocked her head in both directions, looking from part to part with furrowed eyebrows and a slight shake of her head. Re-opening her eyes, she said that she couldn't, because neither really trusted the other and that neither could understand what the other was saying.

The author then instructed her to go back inside and have her therapist part tell the two conflicting parts about how two individuals, just like them, were able to get together, to learn how to communicate, and to cooperate with one another. "And as you do this I want you to allow your hands to begin to rise up toward your face only as fast as you communicate the metaphor to them so that they will fold beneath your chin at the moment the two parts make an agreement in the story."

In this way the author was making use of two sets of anchors to assure a strong integration.

She closed her eyes and shortly her hands began to move toward each other. Her face flattened and her breathing became almost imperceptible. As her hands moved they assumed a slow, jerky movement characteristic of idea motor movement observed in hypnotic trances.

When the hands reached a point about two inches from each other they hesitated momentarily. When the hands came together they clenched tightly. Her head dropped and she began to cry in short sobs. She pulled her hands tightly against her chest and then dropped them into her lap relaxed but still folded. She continued, crying for a few minutes, and then lapsed into a quiet contemplation.

She was then asked if any of the parts had anything more to communicate. Her head immediately began to shake "no," and she said that the two parts were no longer separate. They were the same part. To this she added, in a different tonality as more tears began to form, "I had to work so hard to keep them apart for so long."

She was asked if she would relate the content of her metaphor to the group.

R agreed, beginning by saying that she hadn't thought about the story at all and that it had just spontaneously come as she talked to the two parts.

"The story was about two sisters: one fat and sluggish, the other thin and aggressive. Their personalities were so different that they had managed to grow up having very little to do with one another. They rarely paid any attention to each other. They each stayed in separate rooms, ate at different times and had different friends, with the exception of one very close and gentle friend to both of them. But whenever he came to visit, one of the sisters would leave the room.

"One day the two sisters happened to be out and walking down the same street but in opposite directions, It was one of the fat sister's few trips out of the house.

"They were purposely ignoring each other when they suddenly heard a crash. They both ran to the scene of the accident and discovered it was their mutual friend who had swerved to miss an animal and wrecked his car. He was severely injured.

"The fat sister immediately began to comfort the friend. The other rushed to contact the ambulance and other authorities. Help arrived and the friend was rushed to the hospital. As the two sisters stood by the wreckage they looked at one another and realized that they had probably saved his life, each performing an important and necessary function. As they realized this they embraced and became one person. That's when my hands came together."

Asked if she felt she could repeat the process if necessary, R said, "Yes, but I don't think I'll need to."

As the group continued, R said that she felt as though she had just come out of deep trance. The feeling remained with her until the end of the session.

When she returned the next week, R was very lively and said that her problems with her "anxiety" and her "heaviness" had cleared up. During the training session she related a number of instances where she had successfully used the process of metaphor in her work with her clients.

Appendix B

NEUROLOGY OF LEARNING

A response is anchored to stimulus due to interference patterns and feedback within neural networks that act in a way similar to the holographic process. E. Roy John (1975), who has done much research on electrical brain responses in cats during learning, provides a good deal of insight into this process of establishing a behavioral pattern or program. John identifies two measurable classes of electrical brain activity: (1) *stimulus-bound* activity, generated by external (exogenous) sources; and (2) *emitted potentials* that are internally generated (endogenous) responses evoked by the stimulus-bound sensory input.

John states that during conditioning (the establishment of an anchor) there is an initial widespread "irradiation" of the stimulus-bound sensory information, caused by the external-conditioning stimulus, over many different parts of the brain. (These are recorded as similar high-frequency, low-voltage wave forms, found in the different brain areas.) As the stimulus becomes paired with the response, the high-frequency activity becomes localized to only a few "relevant" brain areas, and is replaced by synchronized low-frequency, high-voltage electrical activity. John writes:

> The critical event, in learning, is envisaged as the establishment of representational systems [*not the same meaning as in NLP—R.D.*] of large numbers of neurons in different parts of the brain, whose activity has been affected in a coordinated way by the spatiotemporal characteristics of the stimuli present during a learning experience. The coherent pattern of discharge of neurons in these regions spreads to numerous other regions of the brain. Sustained transactions of activity between participating cells permit rapid interaction among all regions affected by the incoming sequence of stimuli as well as the subsequent spread. This initiates the development of a *common mode of activity,* a temporal pattern which is coherent across various regions specified for that stimulus complex. As this common mode of activity is sustained, certain changes are presumed to take place . . . , (that) increase the probability of recurrence of that coherent pattern in the network.
> . . . The same ensemble can represent many different items, each with a different coherent pattern . . . new responses are based upon the establishment of new temporal patterns of ensemble activity, rather than on the elaboration of new pathways or connections. Learning increases the probability that particular temporal patterns will occur in couple ensembles of neurons. By this process, the representational system acquires the capability of releasing the specified common mode of activity as a whole if some significant portion of the system enters the appropriate mode.

The "common mode of activity" is established through the modification of synaptic and neural-membrane potentials. Pribram writes that the pairing of the stimulus and response

> . . . engages the junctional and dendritic mechanisms of the brain where the slow potential microstructure, the holo-

graphic representation of input, is produced. Only with repetition do patterns of these slow potentials intercorrelate sufficiently to generate the nerve impulses necessary to action. Each slow potential pattern is assumed to leave its residue at these synaptic junctions and dendritic locations and so participate in generating the correlations.

A behavioral program is the result of a common mode of neural activity (established through cortico-reticular feedback and synaptic changes) that arises in the central cortico-reticular region and propagates to other brain regions in a systematic sequence. The coherence of these emitted potentials is not in the particular set of neurons they activate (as with the stimulus-bound activity), but in their firing rhythm. These rhythms are originally established in response to an external stimulus (in John's conditioning experiments), but eventually consolidate into coherent slow-wave brain states capable of occurring as a whole when "some significant portion of the system enters into the appropriate mode." John notes that the emitted pattern associated with the learned behavior, although absent in the animal's home cage, began to appear as soon as the animal entered the training environment. The animal began to respond to other anchors established by the training environment.

Further, John found that he could produce an electrical facsimile of the rhythm of the emitted potential by recording their wave shapes through electrodes implanted in the reticular formations of the cat's brain. He then played back the recorded potentials and was able to elicit the conditioned response, even though the appropriate sensory cue had not been given. John then recorded the emitted potentials for conditioned responses to varying frequencies of visual and auditory cues. These emitted patterns were then pitted against concurrent discordant visual cues of other frequencies. John reports that in every case the played-back emitted pattern "completely or substantially controlled the behavioral outcome, preempting the decision and effectively contradicting the concurrent auditory or visual cue."

Of direct importance to the understanding of NLP are the experiments John did involving the interruption of established, emitted potentials. He found that it was possible to interrupt stimulus-controlled behaviors by electrical stimulation of a wide variety of brain regions. He reports:

> Electrical stimulation of the sensory cortex during the exogenous components produced no disruption of discriminative responses, while stimulation with identical current parameters timed to coincide with endogenous processes [*that is, after the animal had begun its behavioral program—R.D.*] ... totally abolished conditioned responses.... In many cases, this occlusion persists for many seconds or even a few minutes after the termination of the electrical stimulation. Recording during this 'post-stimulus absence' reveals high-voltage spindle waves in the intralaminar nuclei, independent of the locus of the electrical stimulation which produced the absence. Seizure-like afterdischarge can be produced in a variety of regions, especially in the limbic system, by such a procedure.

To reiterate the important points of this research: (1) A behavioral pattern (often called a TOTE in NLP), represented by a particular 6-tuple, is the result of an orderly firing pattern, in extensive neural ensembles, that propagates from lower brain structures and establishes itself through a feedback process as a cortico-thalamic resonance (probably similar to a capacitor and inductor in an electric circuit) and is measured as a slow-wave potential. Due to changes made in synaptic and dendritic connections, the whole resonance rhythm can be reproduced only when parts of the system have entered the appropriate mode (a process similar to the optical hologram).

Note: It is obvious that the structural properties of the neural networks involved in the resonance (such as the development of primary representational system and lead system properties as the brain of an individual matures) will have a great effect on the final behavioral outcome of the individual. They will, undoubtedly, determine the types of resonant rhythms, the significant brain regions involved, and the amount of electric "signal" within particular brain regions for any specific emitted potential. The parameters of the 6-tuple (A,V,K_t,K_v,O,M) remain constant. All experience is represented by the same perceptual structures. Individual programs/6-tuples are made by different firing rhythms, signal displacement, and other syntactic parameters. John writes that an ambiguous stimulus activates a variety of emitted potentials and that "the behavior eventually displayed seems to depend upon the particular readout mode [*the emitted pattern—R.D.*] which becomes dominant. . . ." There is no doubt that these types of multiple responses constitute *parts*.

Another interesting insight into the neurological representation of parts and their integration is the work of Morrell et al. (1967), cited by John, that showed neural cells responsive to two different stimuli displayed an altered pattern of activity when the two stimuli occurred together (when the anchors were presented simultaneously). Subsequently, presentation of one of the stimuli alone elicited the pattern seen during pairing.

2) Internally propagated electrical activity can preempt activity initiated by external cues.

3) The interruption of a slow-wave emitted potential produces high voltage spindles in lower-brain structures, regardless of the origin of the interruption, that abolish conditioned responses and produce a seizure-like afterdischarge (that is, a somnambulistic trance).

A final point, important to this discussion, is that the development and consolidation of slow potentials occurs as the individual becomes overconditioned to the response: that is, as the response becomes more and more habitual and automatic. Pribram (1976) correlates conscious awareness directly with the development of habitual behavior and the concurrent establishment of slow potentials. The slow potentials do not supply enough signal to be conscious. It is only the high-frequency activity that, as John claims, becomes localized to a few relevant brain areas that retains enough signal to be conscious. Therefore, those programs that have become the most established are the least conscious.

Given the above information one can predict that, by supplying a sufficiently strong input to the cortex of another individual perceptually through some overt behavior, a programmer can produce results similar to those recorded by E. Roy John in his pattern-interruption experiments. An account of such a procedure is given in this paper.

Since those programs that are most overconditioned and unconscious will have the most consolidated slow potentials, interruption of these programs will be the most profound. Behaviors of this type would include those established at an early age, such as language, accessing cues, and other mannerisms, much social behavior (such as handshakes), and rote skills.

Because the interruption state is so unique, there is minimal interference from other anchors to the holographic process, and new programs may be established rapidly in the new state. Further, the programs will be unique to this state because the new emitted potentials will be established in conjunction with the activity defining the interrupted state. An additional consideration is that since the continuity of an individual's behavior has been interrupted, he/she is momentarily left without a next step for his/her behavior. The programmer may then supply an anchor, easily tracking the individual to a desired response.

Again, the programmer should be cautious with this tactic, as the high-voltage spindles in the limbic system are often subjectively experienced as unpleasant visceral sensations. The adverse subjective feelings are an adaptive response to the interference, or breakup, of an individual's reality.

BIBLIOGRAPHY

Bandler, R., Grinder, J.: *The Structure of Magic I & II.* Science and Behavior Books, 1975, 1976.

Bateson, G.: *Steps to an Ecology of Mind.* Ballantine Books, 1972.

Callaway, E.: Schizophrenia and Interference. *Arch Gen Psychiat* 22: 193-208 March 1970.

John, E. R.: *A Model of Consciousness.* 1975.

Miller, G.: The Magical Number Seven, Plus or Minus Two. *Psych Review* 83: 81-97, 1957.

Galin, D., Ornstein, R.: Individual Differences in Cognitive Style—Reflective Eye Movements. *Neuropsychologia,* 12, 397-376, 1974.

Dilts, R.; "EEG and Representational Systems," 1977.

Dumas, R., Morgan, A.: EEG Asymmetry as a Function of Occupation, Task and Task Difficulty. *Neuropsychologia* 13: 219-228, 1975.

Kocel, K., et al.: Lateral Eye Movement and Cognitive Mode. *Psychon Sci.* 27: 223-224, 1972.

Kinsbourne, M.: Eye and Head Turning Indicates Cerebral Lateralization. *Science* 179: 539-541, 1972.

Grinder, J., Bandler, R., DeLozier, J.: *Patterns of the Hypnotic Techniques of Milton H. Erickson, M.D. Vol. I and II.* Meta Publications, 1975, 1976.

Pribram, K.: *Languages of the Brain.* Prentice-Hall, 1971.

———; The Holographic Hypothesis of Memory Structure in Brain Function and Perception. 1974.

———; Problems Concerning the Structure of Consciousness. *Consciousness and the Brain*, 1976.

Ashby, W.R.: *Design for a Brain: The Origin of Adaptive Behavior.* John Wiley & Sons, 1960.

———; *Introduction to Cybernetics.* London University Paperbacks, 1964.

Sheldon, W.H.: *The Varieties of Human Physique; An Introduction to Constitutional Psychology.* Hafner, 1940.

———: *Varieties of Human Temperament; A Psychology of Constitutional Differences.* Harper, 1942.

———: *Atlas of Men; A Guide for Somatyping the Adult Male at All Ages,* Harper, 1954.

John, E.R.: "Switchboard vs. Statistical Theories of Learning and Memory," *Science* 177. 850–864, 8 Sept. 1972.

Morrell, F. et al.: "Electroencephalography. *Clinical Neurophysiology* Vol 23, 1967.

Haley, J. (ed.): *Advanced Techniques of Hypnosis and Therapy.* Grune & Stratton 1967.

———; *Uncommon Therapy.* W.W. Norton & Co., 1973.

Thomason, Arbuckle, and Cody; "Test of the Eye Movement Hypothesis of Neuro-Linguistic Programming," *Perceptual and Motor Skills* Vol 41 p. 230, 1980.

Owens, L.; "Eye Movements and Representational Systems," 1977.

www.ingramcontent.com/pod-product-compliance
Lightning Source LLC
Chambersburg PA
CBHW071727080526
44588CB00013B/1922